DREAMS, DEATH, REBIRTH

Dreams, Death, Rebirth

*A Topological Odyssey into Alchemy's
Hidden Dimensions*

STEVEN M. ROSEN

CHIRON PUBLICATIONS
ASHEVILLE, NORTH CAROLINA

www.ChironPublications.com

Re-released in 2015 under ISBN 978-1-63051-279-8 for distribution purposes.

Passages in the text are reprinted from *Topologies of the Flesh*, © 2006 by Steven M. Rosen. This material is used by permission of Ohio University Press, www.ohioswallow.com.

Book and cover design by Marianne Jankowski.
Printed in the United States of America.

978-1-63051-279-8 (paperback)
978-1-63051-348-1 (clothbound)

Library of Congress Cataloging-in-Publication Data
Rosen, Steven M.

    Dreams, death, rebirth : a topological odyssey into alchemy's hidden dimensions / Steven M. Rosen.
      pages cm
    Includes bibliographical references and index.
    ISBN 978-1-63051-083-1 (pbk. : alk. paper) -- ISBN 978-1-63051-084-8 (hardcover : alk. paper)
    1. Death. 2. Identity (Psychology) 3. Alchemy. 4. Topology. 5. Conciousness.
    I. Title.
    BD444.R675 2014
    128'.5--dc23

                                                               2014021597

To Jonas and Hannah, Jared and Gus,
To my soul-sister Geo,
And to my *anima mundi*, Marlene

CONTENTS

ILLUSTRATIONS

FOREWORD

Depth psychology began with Sigmund Freud's discovery of the nature of an unconscious layer of existence affecting what we imagine, think, and do. Psychoanalysis eventually moved beyond Freud's belief that repression of the oedipal complex and the energies of infantile sexuality create the symptoms and character distortions of mind and body dominating our culture. This approach shifted to addressing omnipotence as the core issue, which, in turn, was supplanted by the analysis of a patient's psychotic features.

As Freudian-oriented analysts continued their explorations of personal or developmental aspects of character and identity, related schools of thought arose. The work of Melanie Klein was exceptional in its designation of "paranoid-schizoid" and "depressive" positions, both of which precede the oedipal stage. Wilfred Bion deepened Klein's insights with his emphasis on a spiritual dimension he called "O" and the interrelation between the two positions she discovered, while neo-Kleinians traced psychic origins into prenatal life.

With these attempts at understanding the most troubling conditions, such as autism, borderline states, and psychosis, psychotherapy concentrated on "object relations," especially after the once-believed impossibility of analyzing narcissism was breached by Heinz Kohut's work, which he named *self psychology*. In addition, schools of thought emerged that dealt exclusively with the *relation* between analyst and analysand. And alongside these developments, much like a dissociated "alter" with which a great number of analysts in the aforementioned areas have minimal acquaintance, is the contribution of C. G. Jung, further developed by post-Jungians.

With Jung we meet an attitude insisting that personal history, no matter how far back it is taken, cannot on its own do justice to human experience. Instead, along with personal, historical causes—themselves given either prominence or marginal notice, depending on the particular post-Jungian predilection—the concept of an archetypal, or transpersonal, dimension is dominant.

The variations among all these approaches, Jungian and others, are broad in scope and have yielded a Tower of Babel of terms, many of which, while going by the same name, such as *self*, mean different things to different analysts. Within this mass of confusion entered psychologists with a scientific focus on "outcome," and this has led to widespread loss of faith in the entire psychotherapy enterprise and to the dominance of so-called cognitive and short-term, strictly symptom-oriented therapies.

The endeavor, begun with Freud, has lost its shine; psychotherapy, with its time demands and expense, and its often dubious outcome, is now of only limited appeal, at least relative to its heyday in the 1950s, '60s, and '70s. Nevertheless, psychoanalytical discoveries of the last one hundred years have been culturally deeply enriching and have significantly helped thousands of people. It is premature to count out a large-scale reawakening of depth psychology.

There are, I believe, relatively unexplored reasons for the decline. One topic to be considered is the kind of consciousness necessary to deal with the psyche in a manner that arouses its own healing potentials. This requires a fundamental change from the rational-scientific, subject-object mode of consciousness. In this regard, a major candidate is the aperspectival form of awareness described by Jean Gebser.[1] An associated and pivotal theme is that of containment. How can psychophysical life be contained in a transformational way, rather than generating new forms of repression or sublimation?

Is such containment created, for example, by the analyst's empathy, or by the capacity to take in, through imagination, the patient's psychic contents and to reflect on these "induced reactions"? Or is the issue of container more complicated, requiring a world of insides and outsides with more than three dimensions? Depth psychology, with some notable exceptions (perhaps found implicitly in the writings of Bion and Winnicott), has rarely entered into these areas. It is here that the trailblazing work of Steven Rosen contributes to our understanding. His explorations of the topology of the *vas hermeticum*, the container of alchemical transformation, bring to light the essence of containment in psychotherapy.

Rosen's research began with topological investigations of the Moebius strip, and, most importantly, the Klein bottle, that topological entity in which one can feel situated inside the vessel only to then oscillate quite naturally onto its outer surface. The enigmatic Klein bottle requires four dimensions for its proper construction, and in this extra-dimensional realm, inside and outside are the same surface.

Here is an example that demonstrates the meaning of the Klein bottle as a container of psychic life.

A woman's friend asks a favor. While she is in town over the weekend, could she stay in the woman's office apartment? The woman felt disturbed by the request. Even though she was not using the office during the proposed time, her friend's wish seemed intrusive. In the past, her knee-jerk reaction would have been to be generous and agree, but this time feelings of discomfort could not be readily pushed aside, and she told her friend that staying at her office wouldn't work.

Soon afterward she felt a growing pain in her chest and stomach, and then a compulsion to call the friend and say circumstances had changed and it was now okay. She restrained herself, not wanting to fail in her resolve. But the pain was surprisingly strong, and she thought agreeing to the request would surely lessen it.

In therapy, she had often focused on her near-automatic behavior of being generous as a way of avoiding traumatic feelings. In this instance she was able to stop going over in her mind whether agreeing or not agreeing to the request was right. Nor did she let herself obsess about the friend and her personality issues in a negative, distancing way. None of this relieved her pain. However, she found the courage to actively stay with her subjective condition, not to *do* something with it, such as analyze it, portray the feelings by painting them, or talk to a younger part of herself—all tricks of the trade of various psychotherapy procedures.

Instead, by totally *being* with herself in her pain and near panic, there was a remarkable turn: she experienced a shift to observing herself, as though she was now looking at herself, hunched over in pain and suffering. Then, in a minute or two, there was another shift: she was inside what felt like a container or vessel, undergoing the confused and tormenting feelings and bodily pain. She allowed the oscillation between inside and outside to continue. On one occasion, when she was outside the vessel she could ask herself, inside: "What happened to you?" And "she" replied, "there is a demon in me."

After a number of such oscillations, something totally unexpected happened: she no longer felt the extreme pain and emotional distress. It was not that the demon had vanished, but rather it had transformed into something manageable and she could easily stay with her resolve not to give in to her friend. She felt relieved at not having acted falsely by becoming compliant. This was the first time she had found a container for the anguish that had plagued her life.

This woman had experienced the mysterious topology of the Klein bottle, the alchemical *vas hermeticum*. The spontaneous oscillation between inside and outside, a dynamic that seemed autonomous, not something she imposed on herself, proved transformative. The newly found container extended to other life situations in which she normally would have been locked in a state of fusion with an inner or outer condition, say, an emotion of anger or another person's desires, unable to stay connected to the object or to separate and have her own point of view.

The Klein bottle has a hole in it; it is not a completely continuous surface, and one feels resistance to this jump or discontinuity. Overcoming resistance illustrates the role of subjectivity, so central in Rosen's analysis, in engaging the discontinuity of the Klein bottle.[2] And part of this woman's experience involved the choice to allow and encourage the oscillation, rather than to stay stuck in the familiar territory of suffering or in a disembodied detachment.

The particular condition she encountered has a mythical form in the well over 2,000-year-old story of Attis and Cybele. The boy Attis can neither stay with the Mother Goddess Cybele nor can he separate without the disastrous consequence of going mad. (I have conceptualized this story, from the standpoint of the rational ego in its three-dimensional world, as the Fusional Complex.)[3] The woman I am discussing endured this "impossible state," in which one can neither separate from an object nor stay connected to it, leading to a mind-wrenching suspension in limbo, which also engendered the intense physical and emotional pain she felt. This condition always resists heroic, overcoming efforts, but with courage, one can become active enough to consciously confront the awful, stuck state and the pain.

Jung, in a remarkable intuition, recognized the Attis-Cybele myth as the *prima materia* of alchemy. But transformation of its impossible fusional state can only occur if there is a container for the condition, namely, the four-dimensional Klein bottle. Humanity today still suffers from the Attis-Cybele myth because its dynamics cannot be contained as something "inside," or through seeing its dynamics as induced by another person "outside." Science's tactic of bringing order to phenomena by objectifying them fails here.

Where is the oscillation and transformation occurring? Is it in the mind or in the body? Neither one nor the other fits experience. Instead the changes in states of mind and body take place in the in-between world of the *subtle body*, a concept discarded by science long ago. As

Rosen emphasizes, the subtle body has the geometry of the Klein bottle. Such mainstays of centuries of thought, before science and rational-discursive thinking, do not fit into models with insides and outsides, or into analysis of cause and effect.

It is essential that we recover such previous categories of thought and imagination which have been distorted or completely cast out by rational dominance. In the terminology of historian Ioan Couliano, science uses its selective will to interpose an interpretive grille between preexisting contents and their modern treatment. The result, he says, is a systematic distortion or even a semantic inversion of ideas, which pass through the interpretive grille of the era.[4]

Steven Rosen's body of work (especially *Dimensions of Apeiron, The Self-Evolving Cosmos,* and *Topologies of Flesh*) reclaims these vital ideas. His writings fulfill the spirit of Sandor Ferenczi's notion that psychoanalysis must return to a purified animism. This was a penetrating intuition at the time (1926), but it was flawed by being cast in the three-dimensional, rational framework of psychoanalysis. And clearly, as Rosen so adeptly implies, a "purified animism" would be only one component (the "magical" aspect) of the integral consciousness required.[5] In general, Rosen re-visions alchemical transformation, through new ways of thinking and associated ontology, as the means by which current thought and culture must advance.

In *Dreams, Death, Rebirth,* Rosen takes on that most difficult and complex subject, the death of an ego-centered life and the birth of a Self-centered one. And more: his goal is creating the necessary, topological mode of containment for integrating the body as part of that transformation.

Thus, in this book we find an evolution of Rosen's thought that engages the mystery of death. Here we further discover the complexities of the need for the author to make himself transparent, and for a departure from objectivity into the (carefully thought out) embrace of subjectivity.

All of Steven Rosen's work demonstrates themes of containment and subjectivity, but most fully in *Dreams, Death, Rebirth.* It is a remarkable achievement, and any psychotherapist intent on psychic depth and mysterious regions far beyond rationality, especially the nature of the in-between space joining therapist and patient, will be enriched by it, as will the entire field of psychotherapy.

Nathan Schwartz-Salant
February 2014

PREFACE

Our greatest certainty and greatest mystery is our mortality. In this book, I explore the subject of death and rebirth from a philosophical perspective that addresses the question of human identity. What death and its possible survival mean to us depends on the meaning of "us"—they depend, that is, on what we assume to be the boundaries and limits of our being. In the chapters that follow, I will show how our sense of who and what we are has evolved through the millennia, and we will see what this implies for our understanding of death and the prospect of rebirth.

This will not be a purely academic account. The philosophical and historical narrative is interwoven with my personal experience, particularly through the medium of my dreams, whose contents serve to mirror the themes of the text in a concretely embodied way—like a "Greek chorus" echoing forth from the unconscious. The intimate connection between dreams and death was brought to light by archetypal psychologist James Hillman. In *The Dream and the Underworld*, Hillman associated the oneiric realm with the realm of the dead: "If each dream is a step into the underworld, then remembering a dream is a recollection of death and opens a frightening crevice under our feet."[1] The dreams I recall in this book can be seen accordingly as excursions into the netherworld.

The present volume carries forward work I did in my 2006 book, *Topologies of the Flesh*.[2] I ended *Topologies* by taking some tentative steps into the subterranean sphere, guided by the ancient discipline of *alchemy*. Though alchemy is commonly regarded as an absurd flirtation with transmuting cheap metals into gold, C. G. Jung discovered that it actually involved a sustained and serious effort at transforming more than just matter, but the human psyche as well.[3] And alchemical transformation entails nothing less than a confrontation with death that aims at rebirth.

In view of that, the discipline of alchemy will provide the framework for the present investigation.

My previous work with topology will also be taken further in the pages ahead.[4] Topology is the branch of mathematics conventionally concerned with the properties of geometric figures that stay the same when the figures are stretched or deformed. While mathematics is often a highly abstract enterprise and topology has certainly followed suit, the way I am going to use it, the contrary will be true. I will be proceeding more in the spirit of philosopher Maxine Sheets-Johnstone, who spoke of topology as being "rooted in the body" (she worked with the idea that stretching and twisting our bodies are essentially topological operations).[5] And I will be guided by communications researcher Anita Hammer, who noted that, when a new direction for thinking is indicated, topological imagination can give it concrete form.[6] Indeed, it is concrete imagination that is needed for alchemy's paradoxical encounter with the underworld, and in the journey I am planning, the use of certain odd topological structures will facilitate the imaginal process.

The first chapter of the book opens with a dream that foreshadows the theme of death and rebirth. I then proceed to examine the relationship between mortality and identity in a historical context going back to ancient culture. In chapter 2, the crisis besetting modern culture is acknowledged. The grave challenges confronting humankind are explored as they relate to the denial of death. We discover the need for human awareness to move backward to its unconscious roots where it encounters the *uroboros*—the tail-eating serpent that is perhaps the most primordial symbol of death and rebirth. This archaic figure is also a key symbol of alchemy, and the alchemical approach is seen as crucial in responding to humanity's dilemma. In chapter 3, we follow the call of some contemporary thinkers for a revival of the old practice of alchemy in a new form. Enter here topology. Alchemical imagination is enhanced by enlisting topological structures such as the Moebius strip and Klein bottle—structures that bear a remarkable resemblance to the enigmatic vessel in which the work of alchemy was said to have been done.

Each of the four remaining chapters of the book concerns itself with a different stage of the alchemical opus, as initially described by Jung.[7] In each case, the alchemist's aim is to achieve a *coniunctio* (conjunction): to integrate consciousness and deepen it, bringing awareness to primary spheres of knowing and being. In moving successively through the last four chapters, the focus of the work shifts from one form of psychological functioning to another (thinking→feeling→ sensing→intuition); one stage of development to another (adult→ child→infant→embryo); one epoch of cultural history to another (mental-rational→mythic→magical→archaic); one organ system of the body to another (brain→heart→groin→gut); one order of phylogeny to another (human→animal→vegetable→mineral); and so forth.[8] And each chapter reaches its climax in an alchemical encounter with death that opens up the possibility of a transformative rebirth. In the most general terms, movement through the final four chapters of the book traces the path of consciousness backward and downward to earlier, more primordial, more concretely embodied strata of psyche and world. This retrograde passage of awareness is the crux of the alchemical odyssey—a journey that I believe can be helpful in attempting to address the profound global dilemma we currently face.

The work of alchemy can be described in another way, through its ancient motto: *solve et coagula*, "dissolve and coagulate." The alchemical substance to be transformed was first dissolved, broken into parts, sublimated so as to refine it. Then, in the second phase, the process was reversed and the material was made to congeal, its constituent elements being drawn together and solidified. This was the stage of "coagulation." So what had been taken apart was now put back together in a refined form.

Alchemy's old motto applies to this neo-alchemical text, which incorporates both a *solutio* and a *coagulatio*. In portions of the present volume, the subject matter is "taken apart," that is, analyzed and approached abstractly.[9] This is the conventional way of working with the modern text, that wherein the author remains largely detached and anonymous. Of course, the dreams and personal reflections I have brought into the book are part of the *coagulatio*. Note that some of the dreams recounted here are given in the body of the text, while others are gathered in dream journals appearing at the ends of the chapters. All dreams are set in italics, and in the dream journals they are sometimes preceded and/or followed by commentary. Dreams in a journal are linked to particular words or phrases found in the main text. These terms are marked by the special superscript [D], followed by the number of the dream. The first such dream in a chapter is thus marked [D1], the second [D2], and so on. So, in reading the text, when you see the special superscript, you are invited to consult the journal at the end of the chapter for the corresponding dream.

Another aspect of the *coagulatio* involves fleshing out the "delivery system" for the ideas presented. That is to say, if the material composition of the text consisted merely of these typographic marks you are reading—these arbitrarily devised, conventionally agreed upon, one-dimensional tokens—my abstract words about concretizing the text would fall far short of becoming the living reality they signify. As a step toward bringing the physical signifiers of the text into correspondence with what the text signifies, I have added a number of vivid images from alchemy and topology—including a unique stereographic image that conveys a tangible sense of three-dimensionality (the *dimensionality* of the signifier proves to be a critical factor in coagulating the text). Included as well are some photographs and drawings that have much personal significance for me.

In the closing sections of the last four chapters, I attempt a more concrete kind of self-reflection than even personal experience allows. Here, the author of the text does not just make his presence felt in a personal way. Through a certain form of meditation, an opening is created for the possibility that the author's particular self may become aligned with a universal or transpersonal Self (the Self that Jung wrote of in his volumes on alchemy).[10] We will see that this alignment is part of the retrograde movement of consciousness described above, and that the full-fledged concretion of the text depends on said alignment. Only through the "joint authorship" of self and Self can the textual *coagulatio* be realized in full.

ACKNOWLEDGMENTS

I want to thank a number of people for the encouragement and support they have given me as I worked on preparing this book. I am particularly indebted to John Dotson (www.acharantos.com), Lloyd Gilden, Wally Glickman, Deborah Hillman, Anthony Judge (www.laetusinpraesens.org), David and Cyndy Roomy, and Ernie Sherman. My gratitude also goes to David and Jana Dichelle, Marketa Goetz-Stankiewicz, Joel Kroeker, Lael McCall, Lisa Maroski, Adair Linn Nagata, Derek Robinson, Plamen Simeonov, Josh and Lesley Sullivan, and John R. Wikse. For their guidance in seeing the work to fruition, I give my thanks to Chiron editor-in-chief Len Cruz and publisher Steve Buser. I am grateful as well to Vanessa Schwartz and Paul Mendelson for their diligent assistance in the production of this book. Serving as "midwife" for the enterprise was project manager Siobhan Drummond, and I much appreciate her responsiveness, patience, and effective dedication to every aspect of the "birthing." I owe special thanks to Nathan Schwartz-Salant: his helpful feedback, sage advice, and steadfast belief in my work have sustained me. Finally, my love goes out to two women who have been there for me from the inception of this project: Geo N. Turner, my cherished sister, and Marlene Schiwy, my beloved wife.

CHAPTER ONE

Identity and Death
Modern and Ancient Views

Dream of September 20, 2005:

I'm traveling in the entourage of a well-known and charismatic philosopher. We are in a large, outdoor area, participating in a public conference attended by many. It's a momentous occasion and there is a feeling of sprawling openness and extension.

At some point in the dream, the philosopher seems to be quite sick and people are attending to him. In fact he is dying and this is indicated by a huge, egg-shaped mass protruding from his abdomen. The massive ovoid distension points downward toward the earth and has the color of a cabbage. There are strands of purple predominating but other colors as well, some white, and a bit of green. And the protrusion seems "pregnant," about to burst open.

I want to help him and feel I can. I ask him if he's ever had this experience before and he smiles and says he has. So the philosopher is in trouble and he appears to be dying, but there's a smile on his face. It has happened to him before and his smile tells me that all is well.

INTRODUCTION

Religiously oriented people commonly believe that death is a transition to an afterlife in which the individual enters another dimension or otherworldly environment of some sort—be it heaven, hell, or some variation on these. And the reply of atheists certain that religious convictions about life after death are wishful or fearful fantasies is: "When you're dead, you're *dead.*" These opposing views seem as polarized as they possibly could be. But, to begin our odyssey, what I want to explore is what these views have in common.

Though conventional believers and disbelievers may argue vehement-
ly about survival of death, they tacitly share the same assumption about
what it is that survives or does not: the *particular individual*. On both
sides of the debate, it is taken for granted that human identity is limited
to a localized center of consciousness that inhabits a discrete body during
life. Then, when death comes, individual consciousness transcends the
body—according to the believer—or consciousness is simply extinguished—according to the skeptic. While this understanding of identity
dominates contemporary thinking, in its strongest form it is actually not
much more than 700 years old.

Beginning around the fourteenth century CE, European culture un-
derwent the revolution we know as the Renaissance. Every aspect of
life was affected by this transformation, as society was reorganized and
sweeping changes were introduced in government, art, music, science,
mathematics, and philosophy. Moreover, the transmutation of cultural
forms and practices reflected deep-seated internal changes. Human consciousness itself was transfigured.

Philosophers David Lavery and Jean Gebser pointed to the fifteenth-
century introduction of perspective in art as indicating an underlying
alteration of perception: the artist now stood completely apart from others and from nature, observing them at a distance from the fixed point
of reference inside his head.[1] Gebser's recognition of the broad implications of perspective led him to speak of the Renaissance as ushering
in a "perspectival world," one involving an "emergent objectifying consciousness" with a "heightening of awareness . . . accompanied by an
increase of personal ego-consciousness."[2] Correlated developments in
the field of written expression have been examined by language theorist Walter Ong.[3] Ong particularly stressed the far-reaching impact on
consciousness of the advent of print (ca. 1450): how this new medium
standardized, objectified, and spatialized language for use by a detached,
abstractly self-possessed individual. Philosopher Martin Heidegger too
wrote of the transformative implications of the Renaissance.[4] Citing
Galileo's description of his own process of discovery—"I think in my
mind of something moveable"—Heidegger interpreted this "I think" as
mirroring a new sense of being, one that autonomously grounds itself
and secures itself without reliance on anything external. Just a generation after Galileo's self-reflection, the novel order of being came to the
fore in philosopher René Descartes's famous *cogito ergo sum*: "I think,
therefore I am." Let us say then that the Renaissance marked the birth
of the sharply focused experience of discrete individuality so familiar to

us today. It is this localized, insular, seemingly self-contained sense of identity that we have come to feel survives death—or does not.

How was human identity experienced prior to the Renaissance? The philosopher Owen Barfield intimated that, in the pre-Renaissance era, "the world was more like a garment men [and women] wore about them than a stage on which they moved.... Compared with us, they felt them-selves and the objects around them and the words that expressed those objects, immersed together in something like a clear lake of... 'mean-ing.'"[5] This is surely not to say that there was no sense of individuality whatsoever in the medieval and ancient worlds that preceded the Renais-sance. Greek civilization in particular, with its introduction of philosoph-ical reflectiveness, clearly did much to advance the growth of individual awareness. Still, the experience of standing sharply apart from the world, standing starkly alone as a discrete center of consciousness, is unique to the post-Renaissance epoch. Before that period, the individual was always embedded in a context, an embodied community of some kind, be it family, church, village, social or vocational organization. This had the effect of keeping soft the boundaries that set a person off from his or her surroundings. As Jungian theorist Nathan Schwartz-Salant put it,

> People had little concept of individualism [prior to the Renais-sance]; instead, they considered themselves to be part of a collec-tive reality and organized their lives through myth and custom. The individual was of little import .... As long as people func-tioned through broad, mythical patterns, the individual ego held no mass appeal or value, and in fact was considered a danger.[6]

But in the aftermath of the Renaissance—and especially over the last century, when the experience of separateness has intensified to the point of becoming one of isolation, alienation, or estrangement (even within nuclear families)—the sense of communal belonging has largely dissi-pated.[D1] (Superscript [D1] refers to the first dream in the dream journal given at the end of this chapter; see the preface for more information on dream formatting.) One can certainly belong to many and varied groups, yet one's concrete sense of *who one is*, of the locus of one's actual physi-cal being, more than ever before lies not so much with any sort of group or collective but with the discretely experienced, skin-encased self that is centered in the head, just behind the eyes.

The implications of this for our attitude toward death are profound. Given the compelling sense of being contained within this particular body of mine, it is very difficult for me to imagine my core identity extending beyond the boundaries of my skin in any literal way. Although I may strongly identify with another person or group and feel emotionally connected to it, the "I" that does the identifying and connecting remains grounded in my discrete body, not in any other. With my identity thus circumscribed, I am constrained to believe that, when my body succumbs, I am extinguished—or else I somehow continue in an afterlife as a disembodied (or quasi-embodied) individual consciousness. But the latter belief does require quite a leap of faith. If, on the other hand, my core sense of being were not tied to a particular body sharply set apart from others, if instead it were grounded communally, then without question the demise of this finite body would not simply mean the end. For the communal body in which I was integrally embedded would continue despite the passing of one of its members. Presumably, that wider sense of identity was indeed experienced prior to the Renaissance. But to what extent was it experienced, and in what period or periods of cultural history?

Moving backward in history from the Renaissance, we do not immediately encounter a transition to full-fledged communal identity but find varying degrees of embedment in a social context. It is very difficult, if not impossible, to specify the exact extent to which particular peoples were communally embedded at particular times, and I will not attempt to do so, since that would take us too far afield. What we do know is that, while pre-Renaissance experience was generally more communally grounded than what came after the Renaissance, there was already a formed sense of individual identity among the ancient Greeks. Can the same be said of the period prior to the emergence of Greek civilization in the first millennium BCE? (In the next chapter, the sense of identity characteristic of medieval Europe will be considered.)

THE TRANSFORMATION OF IDENTITY IN ANCIENT CULTURE

There can be no concern about death if there is no core sense of identity, no circumscribed center of self, ego, or being that would have such a concern. In *The Origins and History of Consciousness*, Jungian theorist Erich Neumann provides a detailed account of cultural development that documents the transition from a diffusely communal sense of identity to a more sharply individualized one. Here we see the movement from being oblivious to death to being preoccupied with it, and finally, to repressing awareness of it.

Sampling from a wide range of mythic cultures—such as the Indian, Babylonian, Cretan, Canaanite, and especially the Egyptian—Neumann sets forth the basic pattern of development he called "centroversion," the process through which consciousness achieves a well-formed and stable ego serving as the center of an individualized identity. At the outset, the embryonic ego is embedded in the dark matrix of the unconscious, an underworld of unfathomable depth intimated by the ancient mythic symbol of the self-devouring serpent or *uroboros* (fig. 1.1).

Figure 1.1 Image of the uroboros adapted from an image appearing in the *Chrysopoeia of Cleopatra*, an ancient Egyptian alchemical text (ca. 100 CE).

The primeval uroboros is everywhere evident in early human culture. "There are pictures of it in the sand paintings of the Navajo Indians and in Giotto; it is found in Egypt, Africa, and India, among the gypsies as an amulet, and in the alchemical texts."[7] "The uroboros," says Neumann,

> is the symbolic self-representation of the dawn state, showing the infancy of both mankind and of the child....It functions as a transpersonal factor that was there as a psychic stage of being before the formation of an ego. Moreover, its reality is re-experienced in every early childhood, and the child's personal experience of this pre-ego stage retraces the old track trodden by humanity.[8]

The uroboric stage of development is "the time of existence in paradise where the psyche has her preworldly abode, the time before the birth of the ego, the time of unconscious envelopment, of swimming in the ocean of the unborn."[9] In this time of oblivion, the embryonic ego is scarcely concerned with the matter of death. Sociologically speaking, "the embedding of the germinal ego in the uroboros corresponds ... to the state in which collective ideas prevailed, and the group and group consciousness were dominant. In this state the ego was not an autonomous, individualized entity with a knowledge, morality, volition, and activity of its own; it functioned solely as part of the group, and the group with its superordinate power was the only real subject."[10] To mythic awareness, however, collective subjectivity was not invested in an assemblage of peers but in higher authorities imbued with transpersonal powers: kings and queens, gods and goddesses. This is evidenced in psychologist Julian Jaynes's corresponding comments on how different mythic awareness was from the rational consciousness later to emerge. Mythic peoples "do not sit down and think out what they do. They have no conscious minds such as we say we have . . . . The beginnings of action are not in conscious plans, reasons, and motives; they are in the actions and speeches of the gods."[11] The nascent human ego thus found itself in a largely passive position, completely open to external forces personified by the deities. And the gods themselves were ruled by Fate (or the Fates), embodied by a mysterious goddess[D2] operating beyond the realm of rational thought ("above the gods is fate, a blind, inscrutable 'will'").[12] What I demonstrated in my book *Dimensions of Apeiron* is that this goddess who rules both mortals and gods is none other than the uroboric Great Mother, the pre-rational force of Mother Nature.

The encircling containment portrayed in various images of the uroboros (fig. 1.2) indeed conveys its function as "great Mother Nature," "Primordial Mother," or "maternal womb."[13] Of course, this motherly containment is deeply paradoxical. Great Mother does not just contain her child as would an ordinary container, a vessel that is distinct from its contents and encloses them in merely an external fashion. In swallowing her own tail, the uroboric serpent is *self*-containing; she therefore contains her offspring as *herself*. And this act of swallowing herself confounds all simple oppositions: inside versus outside, self versus other, masculine versus feminine, life versus death, and so on. I will have more to say about the curious character of uroboric containment as we proceed.

Figure 1.2 Various images of the uroboros. Upper left-hand image courtesy of Swiertz Sebastien (Wikimedia.org), upper right-hand image courtesy of Johann Jaritz (Wikimedia.org); below, image of the uroboros adapted from the *Chrysopoeia of Cleopatra* (see figure 1.1).

Neumann makes it clear that once the child is prepared to leave the womb, once "consciousness begins to turn into self-consciousness, that is, to recognize and discriminate itself as a separate individual ego. . . . feelings of transitoriness and mortality . . . now color the ego's picture of the uroboros, in absolute contrast to the original situation of contentment."[14] So the ego begins to fear its demise just as soon as there *is* a separate ego ready to stand apart from the maternal matrix. "Ego consciousness . . . introduces . . . death into man's life," says Neumann.[15]

The fear of death is the ultimate fear. In the first hesitant steps out of the maternal womb, the greatest danger to the weakly formed ego seems to lie in losing support from the mother with whom it is still primarily identified. But when centroversion advances and the center of identity shifts further away from the mother, the anticipation of death takes form as the fear of being sucked *back into* the uroboric womb where the still-tentative ego would be overwhelmed and dissolved. In the words of Neumann, "the transition from the uroboros to the adolescent stage was characterized by the emergence of fear and the death feeling, because the ego, not yet invested with full authority, felt the supremacy of the

uroboros as an overwhelming danger."[16] This fear of death is so fundamental that whole civilizations were organized largely around trying to
deal with it. Ancient Egypt is the best-known example. What we learn
from Neumann and many others who have written about this culture is
that its elaborate burial rituals and meticulous embalmment procedures
mirrored a pervasive obsession with the task of assuring the immortality
of the departed soul. Neumann sees mythic Egypt as representing a tran-
sitional stage of cultural development from a condition of egolessness to
one in which the ego remains relatively weak and strives above all else
to break the ominous hold of the uroboros.

Neumann stresses that, in the uroboros, life and death are inseparably
entwined: "decay and the necessity of death are one side of the uroboros
just because its other side signifies birth and life."[17] Thus, Neumann—
referring now to another uroborically dominated pre-rational culture,
speaks of "the tendency of Canaanite mythology to bring opposites together, so that . . . the god of death and destruction is also the god of life
and healing, just as the goddess Anat is the destroyer and, at the same
time, the goddess of life and propagation. The uroboric coincidence of
opposites is expressed in this juxtaposition of positive and negative features."[18] Neumann concludes that

> wherever the harmful character of the Great Mother predomi-
> nates or is equal to her positive and creative side, and wherever
> her destructive side . . . appears together with her fruitful womb,
> the uroboros is still operative in the background. In all these cas-
> es, the adolescent stage of the ego has not been overcome, nor
> has the ego yet made itself independent of the unconscious.[19]

In passing from mythic Egypt to rational Greece, "adolescence" is
surpassed. This marks a transformation in the approach to coping with
mortality, one that has been carried forward into our present time: a
stronger, more centroverted ego deals with the subject of death chiefly
by repressing it. This is well reflected in the repression of the uroboros.

Relating centroversion to the rise of male dominance, with its rational,
linear style of operation, Neumann asserts that "patriarchal Greece" re-
pudiates "the hybrid, uroboric state" of the Great Mother.[20] In the Greek
myths, says Neumann, "the terrible aspect of the Great Mother"—that
associated with destruction and death—"is almost wholly repressed."[21]
Commenting further on this theme, Neumann notes: "The growth of
self-consciousness and the strengthening of masculinity thrust the image

of the Great Mother into the background; the patriarchal society splits
it up, and while only the picture of the good Mother is retained in consciousness, her terrible aspect is relegated to the unconscious."[22] We see
clearly the connection with the uroboros when we relate bad Mother or
death Mother to a serpent or dragon. How ubiquitous in world culture is
the patriarchal tale of the male hero winning his freedom and the hand of
the fair princess (good Mother) by slaying the evil dragon. So, "to become
conscious of oneself . . . begins with saying 'no' to the uroboros, to the
Great Mother, to the unconscious."[23] And this means saying no to *death*:
"Stability and indestructibility, the true goals of centroversion," require
"the conquest of death."[24] Today—particularly in the industrialized
West—this repressive vanquishing of death is mirrored in the fact that
the very subject is largely taboo, that dying people are generally kept out
of sight, and that the prospect of our own death may be a source of embar-
rassment and shame that we often deny until the end. Modern culture's
denial of death is famously documented by Ernest Becker, who elaborates
on the myriad ways we ingeniously sidestep the truth of our mortality
and behave as if we will live forever—even though, just beneath the surface of awareness, we are haunted by the specter of the inevitable.[25]

*Journey entry of August 1. For the past four months, I have felt
especially haunted. Almost every day I have been dogged by unexplained
digestive symptoms that include abdominal pain, cramping, and nausea.
In rereading Erich Neumann's book the other day, I noticed that he refers
several times to the "alimentary uroboros." Neumann notes that "the
uroboros is properly called the 'tail-eater,'" and that this paradoxical
symbol of self-digestion dominates both early childhood and the earliest
creation myths—and is then repressed.[26] But this is what I've been
courting and looking to make conscious in the current book project: the
primal circle of eating and being eaten, of life and death. Is it this then that
lies at the core of my digestive distress?*

*What came to mind in rereading Neumann were some potent images I'd
worked with in* Dimensions of Apeiron. *The material in question centers
on a sequence of ominous visions experienced by Zarathustra, a character
in the writings of Friedrich Nietzsche. In the most chilling vision of all,
Zarathustra hears a dog howl in fear under a full moon to announce the
appearance of a man who is lying on the ground: "A young shepherd I saw,
writhing, gagging, in spasms, his face distorted, and a heavy black snake
hung out of his mouth. Had I ever seen so much nausea and pale dread
on one face? He seemed to have been asleep when the snake crawled into*

*his throat, and there bit itself fast."²⁷ It is clear to me that the dreadful
snake stuck in the shepherd's mouth is none other than the "alimentary
uroboros." I was "asleep" when the "serpent" first crawled into my
throat—I was unconscious. To resolve my relentless dyspepsia, it seems I
need to become more conscious.*

*Journal entry of August 21. A week ago I had a CT scan to diagnose my
digestive problems. When I inquired about the results, I was told the scan
showed that an area of my pancreas is "hypodense" (unusually thin).
Internet research then told me that cancer is indicated in 90 percent of
CT scans showing hypodense masses on the pancreas. Suddenly I felt
I'd been given a death sentence. My father died of pancreatic cancer and
now it was my turn. The feeling of doom persisted for much of the day
and into the night, where sleep eluded me. The projection of life opening
on an indefinite horizon into the future had collapsed and my mind
raced in cold claustrophobia. Finally I was able to sleep for a spell after
repeated attempts to drop into my body by bringing my attention to my
breathing.²⁸*

DREAM JOURNAL

🔘 D1

The contemporary sense of isolation was brought home to me personally
in my dream of January 22, 2003:

*I'm living alone on a large farm in a remote rural area. My sense
of isolation has been growing and I feel regretful that I made the
commitment to buy this place. Somehow, I was misled into thinking it was
a good idea, but it isn't turning out that way.*

*At some point early in the dream, I sense that the house is surrounded
by large numbers of dangerous wild animals roaming freely about. Darkly
colored or black, and somewhat featureless, the animals loom in the
background. These ominous creatures frighten me, yet I'm also hoping to
befriend them.*

*Later in the dream, I find myself on the telephone, listening to a
long conversation between unidentified individuals, but I am unable or
unwilling to talk. I'm grateful though, for this minimal contact . . . Still
later, I see people in front of the house. I now have laryngitis and try to*

signal to the others that that is why I can't speak. Smiling apologetically, I
point to my throat.

I do want to communicate. My sense of disconnection has grown to
the point where I'm desperate to break the silence, break through to
the others. When I bought the farm I didn't realize it would have such
a terribly isolating effect on me. There's a powerless feeling of having
less and less of a voice; being less and less able to communicate, as if the
silence has gradually taken over and rendered me mute against my will. I
have the vague sense that, when I first acquired the property, someone had
instructed me on how to work the farm, but I hadn't understood or heeded
the advice.

This dream reflects my own particular experience of isolation now so
prevalent in our culture. When I had the dream nine years ago, I was
writing *Topologies of the Flesh*. That was indeed very solitary work,
and—sitting in my small study day after day for hours on end—I felt
quite cut off from human society. Beyond that, the ideas I was trying to
convey were extremely challenging, leaving me constantly concerned
with how effectively I was communicating them. But what about the
wild animals surrounding my dream farm?

In *Topologies*, I set myself the task of making contact with the world
of animal instinct. It was a matter of bringing to light the "wild animals" within me through a process of self-exploration that scared me
at times, and left me feeling further alienated from the familiar world
of human affairs. The connection between my work on *Topologies* and
the dream became clear when I recalled an experience I'd had the night
before dreaming. I had seen a TV documentary focused on the prehistoric cave drawings of wild animals at Lascaux—drawings I explicitly
deal with in *Topologies*. You will see in the pages that follow that the
realm of animal instinct and emotion continues to be of vital concern
to me.

D2

The rule of the gods over mythic humanity and the potent role of the
goddess in this, seem reflected in my dream of May 18, 2006:

I was part of a large group of people caught up in a series of terrible
catastrophes. The earth was cracking and shifting precipitously, and we
were swept along in massive wrenching movements that we were barely

able to survive. It seemed, though, that this apocalypse was being staged. The figures in control were hovering in the background behind partitions and walls. I could get glimpses of half faces, profiles that were partially concealed. From what I could make out, the shadowy creatures had the strange appearance of two-dimensional cutouts. And these phantoms possessed a scintillating, numinous, archetypal quality that reminded me of the Eleusinian mystery rites.

Despite the turmoil we were experiencing at the hands of the mysterious figures, there were glimmers of hope that we could get through the ordeal, and we finally reached the point where the upheavals began gradually subsiding. The challenges were becoming less taxing and the situation more settled.

Glancing toward the shrouded faces of those in control, I saw the partly covered visage of a stunning woman, and I was suddenly moved to tears. She was radiant amidst the lingering chaos and she looked at me with loving eyes. SHE LOVES ME, I knew. Overcome by emotion, I said: "Oh, my darling, oh, my darling."

All is well. This anima figure, this goddess, is my darling, and I am redeemed!

CHAPTER TWO

Shifting the Gears of Consciousness

THE BACKWARD MOVE

According to Erich Neumann,

> The same uroboric symbolism that stands at the begin-
> ning, before ego development starts, reappears at the
> end, when ego development is replaced by the develop-
> ment of the self, or individuation. When the universal
> principle of opposites no longer predominates, and de-
> vouring or being devoured by the world has ceased to be
> of prime importance, the uroboros symbol will reappear
> as the mandala in the psychology of the adult.[1]

This developmental shift can be seen as applying not just to the development of the individual, of course, but to our entire culture as well.
But have we yet reached the point where "the universal principle of opposites no longer predominates"? I don't think so. While we've surely
reached the point where the polarizations, dualisms, and objectifications
of insular identity are no longer really working for us and have begun
to be challenged, hardly have we overcome the ontological inertia that
has gathered over the centuries to circumscribe being and set it apart.
Old habits are slow to die and *ontological* habits are the slowest of the
lot. "This is what I *am*," says common sense, "an isolated unit of identity set apart from other such units." Still, as the dysfunctional character of the post-Renaissance ego becomes more and more obvious, as the
stresses and strains upon our default identity structure continue to grow,
the long-held ontological habit pattern comes closer and closer to being

disrupted. It is the disruption of that pattern that will enable us to shift effectively into a new gear that *embraces* the uroboros instead of fleeing from it. Embracing the uroboros will mean acknowledging the demise of the insular ego[D1, D2, D3] and consciously immersing ourselves in the communal matrix of the Great Mother that in fact first gives birth to the ego. We are going to see that while this transformation constitutes a death in one sense, in another it entails the *survival* of death.

To be sure, the uroboros is already making its presence felt. But I suggest that, until we are able to recognize it in a fully conscious way, to accept its encompassing nonduality without merely regressing to our undifferentiated mythic past, the uroboric Great Mother will manifest herself solely as a force of destruction. Today we are seeing evidence of this all around us, as we are gripped by fragmentation at every level of human and world affairs: disintegration of families and other social institutions; ethnic conflicts raging around the world; growth in international banditry and terrorism; world markets reaching new levels of erratic fluctuation and economies collapsing; nuclear weapons and waste proliferating out of control; ecosystems strained to the breaking point, unleashing natural catastrophes with devastating consequences (floods, famines, epidemics, earthquakes, tsunamis, hurricanes, etc.). Any one of these problems, taken in isolation, might be attributed solely to specific causes of a local nature. But when we consider the overall pattern, something more fundamental seems to be at play. What I'm proposing is that the chaos surging around the planet reflects the resurgence of the uroboros anticipated by Neumann. Though Neumann might have been thinking that the return of the Great Mother would take a more benign and generative form, perhaps he would have agreed that she would indeed assume her purely destructive, Medusa-like aspect when we attempt to *repress* her reappearance. (My exploration of the return of the *apeiron* parallels Neumann's intimation of the return of the uroboros.)[2]

How can we effectively address the burgeoning chaos now shaking the foundations of our civilization and our very beings? It is *because* our ontologically entrenched habit system is coming undone that we have the freedom to move in a new direction. It is more possible now for us to "switch gears." Into what gear do we need to shift? Into *reverse*, I suggest. That is, instead of our awareness continuing to be projected forward from an insular egoic core, we presently have the latitude of withdrawing that projection, of moving backward into the source of the purportedly isolated kernel of identity. I've termed such a backward movement *proprioception*.

Etymologically, to perceive is to "take hold of" or "take through" (from the Latin *per*, "through," and *capere*, "to take"), and to conceive is to "gather or take in." These activities are carried out in the "forward gear" of consciousness, where attention operates through the objectification of reality carried out by a detached ego. The term *proprioceive* is from the Latin *proprius*, meaning "one's own." Literally, then, proprioception means "taking one's own," which can be read as a taking of self or "self-taking." It is true that the term's conventional meaning derives from physiology, where it signifies an organism's sensitivity to activity in its own muscles, joints, and tendons. But the physicist-philosopher David Bohm spoke of the need for *"proprioceptive thought,"* which he viewed as a meditative act wherein "consciousness . . . [becomes] aware of its own implicate activity, in which its content originates."[3] Another form of proprioceptive practice has been suggested by psychologist/philosopher Eugene Gendlin, who described a method of focusing on psychological issues by drawing attention inward to obtain a *felt sense* of the overall bodily background of the problem.[4] Years earlier, the social psychiatrist Trigant Burrow spoke similarly of the need for human beings to gain a proprioceptive awareness of the organismic basis of their divisive symbolic activity.[5]

It is through the backward movement of proprioception that the projection of the insular ego is withdrawn. All attempts to surpass the ego while continuing to operate in the forward gear must be self-defeating. For, any reflective action the ego may carry out when it is thus oriented is preceded by the *pre*-reflective act of positing itself. Though the ego may consciously intend to transcend or negate itself, these intentions are implicitly undermined, subverted in advance by the very first action it takes: preconsciously, it thrusts itself forward. It is the *I* that wants to negate itself, and this "I" tacitly projected beforehand takes precedence over all conscious intentions. Only by moving attention backward can the ego's forward thrust be counteracted. And in so doing, a glimpse can be gained of the uroboric matrix from which the ego first arises. Through such proprioception, the ego perishes in its familiar form. No longer is it experienced as an isolated center of identity in command of a circumscribed body simply set apart from other such bodies. But neither does the ego simply cease to exist. Instead it is reborn as radically transformed in the womb of the uroboros.

DEATH AND RESURRECTION IN THE ANCIENT WORLD

Jungian analyst Marie-Louise von Franz expresses a related idea in exploring ancient views on death and resurrection. Here von Franz focuses not on the uroboros but on the *unus mundus*, the One World, which, according to C. G. Jung, is "the potential world of the first day of creation, when nothing was yet 'in actu,' i.e., divided into two and many, but was still one."[6] Though she does not refer directly to the uroboros, von Franz does observe that the *unus mundus* is personified by the alchemical figure of Mercurius (Mercury), and there is evidence that Mercurius, for his part, is intimately linked to the uroboros.[7] So, if "Mercurius is most often associated with the uroboros," we may take the uroboros as personifying the *unus mundus*.[8]

Recognizing the *unus mundus* at play in the religions of West Africa, ancient Egypt, and China, von Franz notes that, in each case, "the *unus mundus* is identical with the realm of the dead, the spirit land."[9] In each religion, moreover, the ultimate aim is rebirth after death, which means a transformative awakening of complete self-awareness in the midst of death. African medicine men, for example, venture into the underworld guided by a divinity called "Gba'adu," who "represents the most powerful magic, and 'the highest possible degree of self-knowledge a man can attain.'"[10] Associating Gba'adu with the alchemical Mercurius, von Franz notes that, "according to certain ideas of the alchemists, the individuated human being who has become unified must join himself to this mercurial spirit."[11] This amounts to a reunion with the uroboros (as Neumann foresaw) that must entail an encounter with death:

> In the experience of death the alchemist . . . hoped to discover
> . . . an exalted and final step in the achievement of oneness. . . .
> This experience is particularly beautifully extolled in an ancient
> alchemical text which can be traced back to the Egyptian liturgy
> of the dead. In this text the *unus mundus* is pictured as a transcendental experience of wholeness attainable only in the resurrection mystery after death.[12, D4]

Von Franz explains that the climactic phase of individuation is to be enacted by means of a radical transformation in which the erstwhile rule of the ego is surpassed in realization of the universal Self. But how, specifically, is the domination of the ego to be surmounted in a way that brings the Self to concrete presence?

In seeking to address this question, von Franz is guided by the ancient "Chinese concept of life after death, as described by Richard Wilhelm."[13] Egoic identity is "bound to the physical body"[14]—which means the finite particular body, the body construed as a circumscribed object in space. It is this objectified body that grounds one's ordinary sense of oneself from early infancy. Given this fundamental attachment, the particular ego, operating alone, cannot survive the decimation of the physical body brought on by the ravages of death. Only with the assistance of a *"universal* ego," that is, the Self, is survival possible.[15] To make this generic identity a concrete reality, it is necessary to construct a "subtle body," a "'body of a spiritual kind'": "By building up the spiritual body through meditation exercises, the Chinese attempted in this life to disengage the energies attached to one's ordinary body and thus to endow the seminal power, the entelechy—or, translated into our modern terms, the Self— with a new body."[16] Once this subtle body was fashioned, one was purportedly "able to retain an individual identity after death within the . . . *unus mundus.*"[17]

Importantly, the "meditation exercises" leading to the creation of the "subtle body" involved conscious recognition of a "retrograde movement of life energy."[18] According to von Franz,

> The Self, as a psychophysical monad or ultimate nucleus of the personality, does not merely engender the ego consciousness emanating from it at birth and during the growth of the individual's personality. At death it also draws the ego back into itself and contracts. . . . The moment of death forms the decisive shock . . . as the ego plunges into the inner monad and unites with it. When an individual consciously participates in the individuation process, and thereby prepares himself for this moment by exerting himself to experience it as consciously as he can, he will succeed in experiencing the ego's transposition into the Self knowingly. But when he remains, as it were, hemmed in by floating psychic contents which are autonomous and unintegrated, consciousness becomes deflected and slips into a state of unconsciousness, which the ancient texts symbolized as being imprisoned by underworld demons.[19]

In the terms of Jungian psychology, the meditation enabling the individual to consciously follow the retrograde movement of life energy entails the withdrawal of projection. This is what it means to "integrate

autonomous psychic contents." In the first stages of life, when the movement of energy is "forward," myriad identifications are formed that
serve the development of the particular ego. The contents thus projected
are free-floating or autonomous because the individual is unaware of
the process of projecting them. For the individual to gain the degree of
self-knowledge necessary to sustain her or him in the *unus mundus*, the
projections must be retracted, drawn back in, proprioceived.

However, in the confrontation with death it will not suffice merely
to retract the projections of particular contents, personal identifications
that have hardened into "objective truths" so as to meet the needs of the
particular subject. At a deeper level, it must be recognized that—along
with the objects—*subjectivity as such* has been projected. And—with
the realization of this deeper kind of projection, which we henceforth
call Projection—it is Self or Being per se that is engaged in the backward
movement of Proprioception.

The individuation at stake in the transformative process clearly does
not just involve the development of the particular individual, but of the
Self, of Being as such. We may say, with von Franz, that the Self engenders ego consciousness at birth, and add that it does this in order to
facilitate its *own* development.[20] Then, in the climactic stages of the process, Being reorients itself. It is by shifting into "reverse," by engaging
in Proprioception, that Being endows itself "with a new body," a generic
one that surpasses the finite particularity of the ordinary body.[21] In so
doing, it reaches fulfillment.

Von Franz implicitly deals with this theme of the Self's own development in her interpretation of the Egyptian liturgy of the dead:

> When the One (the Self) acquires its new, adequate, perfect body,
> namely, the supernatural resurrection body; then it destroys its
> former body and permeates all other bodies. Translated into
> modern psychological language this would mean: when the Self,
> in a state of "becoming" within earthly man, fully attains its
> "body," i.e., its goal—the [uroboric] mandala of the *unus mun-
> dus*—it exerts an annihilating effect on man's earthly existence,
> because it has attained a state of being as the all-permeating
> one-continuum.[22]

Let us continue to view the individuation process from the standpoint
of the Self's own development and ask whether Self or Being in fact

could have fully attained its "new body" with the esoteric practices of ancient China and Egypt. In the early stages of promoting its individuation, the Self engenders ego consciousness in such a way that it eclipses itself and the uroboros is overshadowed. Then, when full-fledged maturity is reached, "the uroboros symbol . . . reappear[s] as the mandala in the psychology of the adult," quoting Neumann again. This means that ego consciousness, which had gained sway with the repression of the uroboros, must now face its demise as the dominant factor of the psyche and allow the Self to reemerge to take on its "new body." Was this process completed in the above-mentioned practices carried out by the ancient Chinese and Egyptians? Evidently not. According to Neumann, the Self was still in its "adolescence" at that time and a great deal more ego consciousness had yet to be engendered, for "the ego" had not "yet made itself independent of the unconscious."[23] As cultural philosopher Jean Gebser described the mythic structure of consciousness, it lacked differentiation and was "egoless" (see also Julian Jaynes's related comments in chapter 1).[24] Only in repressing the uroboros and passing from mythic culture to reason-based Greek culture could ego consciousness leave "adolescence" behind and gain a greater measure of autonomy. But the Self was still far from being fully individuated with this initial transition. We are about to see that the process of repression had to be repeated even more forcefully before the uroboros would be able to return to play its role in bringing the Self to fruition.

MEDIEVAL ALCHEMY

During the Middle Ages, the practice of alchemy came to flourish on the European continent. A common misconception is that this hermetic discipline was nothing more than a bizarre flirtation with transmuting base metals into gold. In actuality, the alchemical opus was not just concerned with transforming matter. At bottom it dealt with the transformation of the practitioner him- or herself in the work of individuation. A central image in this work was the uroboros, illustrations of which abound in the alchemical texts (see fig. 2.1; see also Neumann).[25] Did the reappearance of the uroboros signify completion of the individuation process? That was clearly not the case, for individuation had not advanced to the point where the meaning of the uroboros could be contained by the medieval psyche in a fully conscious way. The problem lay in a continuing confusion between the psyche and the material world.

Figure 2.1 Images of the uroboros in alchemical texts. Upper left: from
De Lapide Philosophico (16th century). Upper right: from the *Synosius* (1478).
Below: from *Chrysopoeia of Cleopatra* (ca. 100 CE).

Jung comments on how alchemists saw the relationship between psyche and matter:

> The alchemical *opus* deals in the main not just with chemical
> experiments as such, but with something resembling psychic
> processes expressed in pseudochemical language. . . . In alche-
> my there are two . . . heterogeneous currents flowing side by
> side, which we simply cannot conceive as being compatible.
> Alchemy's "tam ethice quam physice" (as much ethical—i.e.,
> psychological—as physical) is impenetrable to our logic. If the
> alchemist is admittedly using the chemical process only sym-
> bolically, then why does he work in a laboratory with crucibles
> and alembics? And if, as he constantly asserts, he is describing

chemical processes, why distort them past recognition with his mythological symbolisms?[26]

Of course, as Jung well knew, alchemy is "impenetrable to our logic" precisely because our logic separates "hard reality" from that which is "merely symbolic," divides physis from psyche, object from subject. By contrast, the alchemical object is at the same time subject. Jung took pains to bring this out, speaking of alchemy as both a laboratory procedure and a meditation, referring to alchemical "imagination" as "a hybrid phenomenon[D5] . . . half spiritual, half physical."[27] "There was no 'either-or' for that age," says Jung, "but there did exist an intermediate realm between mind and matter. . . . This is the only view that makes sense of alchemical ways of thought, which must otherwise appear nonsensical."[28] The critical point for us is that the uroboric transpermeation of mind and matter found in alchemical thinking during the Middle Ages did not so much reflect a fully conscious fusion of well-differentiated elements as a less-than-conscious *con*fusion based on their non-differentiation.

Art historian Barbara Obrist provides further insight into the medieval mind in her article on the role of visualization in alchemical texts.[29] Obrist points out that alchemical ideation relied largely on metaphor and analogy to convey cosmic and theological truths that were deemed incomprehensible in themselves. This sort of thought process predates the discursive operations that became more influential in the science that followed the Renaissance, where propositional elements assumed to be distinct from one another were linked externally in a causal chain of relations acting locally in a cognitive space ("if X is true and Y is true, then Z must be true"). In the analogical thinking of medieval alchemy, elements were linked *internally*, the association being based on their interchangeability. It was as if, by a kind of "sympathetic magic," the alchemical text—especially the pictorial material (illustrations, diagrams, and tables)—was in direct resonance with invisible truths. Here, no deductive chain of reasoning was believed necessary or possible.

Gebser similarly describes the pre-rational mode of thinking widely operative prior to the Renaissance. This magical mentation is characterized by "analogy or association. . . . Magic man feels things which seem to resemble one another as 'sympathetic to,' or 'sympathizing with,' one another."[30] According to Gebser, magical participation in the world is governed by the principle of *pars pro toto*, where every point of experience is inseparably united with every other point and with the whole. Thus "magic man" inhabits a unitary world "in which each and every thing intertwines and is interchangeable."[31] In such a world—where people "felt themselves and the objects around them and the words that expressed those objects, immersed together in something like a clear lake of . . . 'meaning'" (to quote Barfield again; see chapter 1)—the boundary between inside and outside is porous indeed. This lack of differentiation between psyche and exterior world implicit in medieval ideation made it difficult for the alchemists of this era to achieve their goal of individuation.

Nevertheless, despite the tendency toward magical thinking in medieval alchemy, the subtlety and sophistication of this discipline belie the conclusion that its practitioners were merely unconscious. For one thing, the alchemists well knew that it was of the utmost importance in their efforts at self-transformation that the process be effectively contained. If individuation was to be achieved, the boundaries enclosing the individual could not simply be blurred but would have to be sharply defined. Critical to the aim of proper containment was the alchemical vessel in which the transmutation was to occur.

THE HERMETIC VESSEL

The remarkable work of alchemy was to be carried out in a vessel with remarkable properties in its own right.[32] Jung introduced his discussion of the Hermetic vessel with these words:

> Although an instrument . . . it is no mere piece of apparatus. For the alchemists the vessel is something truly marvelous: a *vas mirabile*. Maria Prophetissa says that the whole secret lies in knowing about the Hermetic vessel. "Unum est vas" (the vessel is one) is emphasized again and again. It must be completely round . . . (the spherical or circular house of glass).[33]

Elsewhere Jung speaks of the "house of the sphere" as the "*vas rotundum*, whose roundness represents the cosmos," this "rotundity" being associated with the realization of wholeness; evidently, the "roundness" must be "simple and perfect."[34] But I suggest that this "perfect roundness" is not fully grasped just by imagining the unbroken surface of a sphere or the circumference of a circle, for, as Jung's studies reveal, the "roundness" of the Hermetic vessel is decidedly *paradoxical* in nature.

One indication of this lies in the fact that the vessel was to be *bene clausum*, well closed or "hermetically sealed."[35] By maintaining the absolute closure of the vessel's surface, the simplicity and perfection of its roundness would be upheld. And yet, the vessel was also thought of as a sieve, an apparatus with openings in it to allow finer substances to pass through.[36] Apparently then, the vessel was to be closed and open at the same time!

The way in which this peculiar requirement could be met becomes clearer when we realize not only that "time and again the alchemists reiterate[d] that the *opus* . . . is a sort of circle like a dragon biting its own tail," but also that the symbol of the uroboros in the form of a dragon or serpent explicitly appears in images of the Hermetic vessel itself (e.g., fig. 2.2).[37] Bearing in mind that this vessel was not merely a piece of equipment to be used in the work of alchemy but was identified with that work in a primary way, we may suppose that the roundness or circularity of the vessel was itself uroboric in character. Engraving the sign of the uroboros on the vessel may therefore be taken to mean that the vessel's very structure is uroboric. Indeed, the surface of such a vessel would not simply be closed as is the surface of a sphere, but open as well. For, while the dragon that has swallowed itself is contained within its own skin (as it would be in a closed vessel), at the same time it is ecstatically *uncon-tained*, that is, beside itself, outside its skin in the open.

Figure 2.2 Uroboros and Hermetic vessel, from *Aurora consurgens*, fifteenth-century treatise on alchemy.

Evidence confirming the ecstatic structure of the Hermetic vessel is found in its association with the symbol of the pelican. Read illustrates a form of the vessel called the "double pelican" (fig. 2.3, left) which "was mystically connected with the process of conjunction [the union of op-posites]."[38] And Jung, in the course of describing the Paracelsan version of alchemical transformation as a *retorta distillatio*, presents another il-lustration of an alchemical container shaped like a pelican (fig. 2.3, right).

According to Jung, the *retorta distillatio* presumably "meant a distillation that was in some way turned back upon itself. It might have taken
place in the vessel called the Pelican where the distillate runs back into
the belly of the retort."[39]

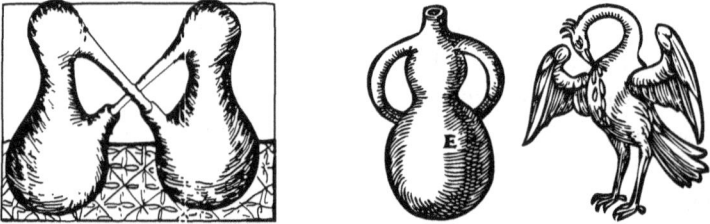

Figure 2.3 Hermetic vessel and pelican. Left: double pelican (from *Das Buch
Zu Distillieren*, Brunschwick, 1519). Right: Hermetic vessel as pelican
(from J. B. Porta, *De Distillatione*, bk. 9, Rome, 1608).

Earlier in the same volume, Jung refers to alchemist Gerard Dorn's
characterization of the Hermetic vessel as the *vas pelicanicum* and further notes:

> The anonymous author of the scholia to the "Tractatus aureus
> Hermetis" says: "This vessel is the true philosophical Pelican,
> and there is none other to be sought for in all the world." It is
> the lapis [the Philosopher's Stone] itself and at the same time
> contains it; that is to say, the self is its own container. This for-
> mulation is borne out by the frequent comparison of the lapis
> to . . . the dragon which devours itself and gives birth to itself.[40]

So the curious roundness of the Hermetic vessel is embodied both in the
uroboros and in the pelican, creatures portrayed as penetrating themselves in such a way that they are inside and outside of themselves at
the same time.

In a chapter devoted to examining the many paradoxes of alchemy,
Jung offers still another perspective on the vessel. According to Jung, the
alchemical paradoxes culminate in the "Enigma of Bologna," which he
characterized as "a perfect paradigm of the method of alchemy in general."[41] On an allegedly ancient monument said to be found near Bologna,
there appeared an inscription that concluded with the following passage:

(This is a tomb that has no body in it.
This is a body that has no tomb round it.
But body and tomb are the same.)[42]

This seemingly nonsensical text attracted much attention among the alchemists, who meditated upon it and devoted great effort to its interpretation. The mysterious persona for whom the inscription was supposed to have been written was named "Aelia Laelia," and according to alchemist Michael Maier, Aelia herself "is the container, converting into herself the contained; and thus she is a tomb or receptacle that has no body or content in it, as was said of Lot's wife, who was her own tomb without a body, and a body without a tomb."[43] Jung identifies the enigmatic tomb of Bologna with the Hermetic vessel. So we witness again the vessel's ecstatic property: outside ("tomb") and inside ("body") permeate each other as one.[D6]

Of particular interest for our purpose is the appearance of the vessel in the material of *glass*. The vessel was referred to as the "house of glass" (as noted above), as a "vessel of diaphanous glass," a "glass vessel that is 'furnished before and behind with eyes' and 'sees the whole universe.'"[44] The symbolic significance of glass is brought out in Jung's association of the Hermetic vessel with the Grimms' fairy tale "The Spirit in the Bottle."[45]

In Jung's view, this story "contains the quintessence and deepest meaning of the Hermetic mystery": a powerful spirit is trapped in the earth beneath an oak tree, enclosed within "a well-sealed glass bottle."[46] Hearing the spirit cry "Let me out!" a passing youth opens the bottle, whereupon the spirit rushes forth, identifies himself as mighty Mercurius, and threatens to strangle his liberator. But the boy tricks the spirit back into the bottle, and the tamed Mercurius then promises that, if freed again, he will serve the boy in a beneficial way.

Jung relates the glass bottle of the Grimms' fairy tale to the Hermetic vessel with the following words:

> The bottle is an artificial human product and thus signifies the intellectual purposefulness and artificiality of the [alchemical] procedure, whose obvious aim is to isolate the spirit from the surrounding medium. As the *vas Hermeticum* of alchemy, it was "hermetically" sealed. . . . It had to be made of glass, and had also to be as round as possible, since it was meant to represent the cosmos in which the earth was created.[47]

In Jung's interpretation, the mercurial spirit represented to the alchemist the initially unconscious, wildly irrational power of instinct, of embodied nature. But Jung's construal of the problem of freeing Mercurius seems less than complete, and I will venture to carry it further.[48]

The fashioning of the Hermetic bottle symbolizes a process of purification which culminates in the *unio mentalis,* a state of intellectual maturity that is the climax of mental development.[49] This is the challenge symbolically faced by the boy in the Grimms' fairy tale, and the challenge the alchemist faced. Prior to "properly sealing the bottle," Mercurius was always a threat to escape, a regression to the primal past that would overwhelm the alchemist. In the meanwhile, the alchemist had to keep the "spirit" imprisoned as best s/he could.

Note Jung's observation that "the alchemists rightly regarded 'mental union in the overcoming of the body' as only the first stage of conjunction or individuation. . . . In general, the alchemists strove for a *total* union of opposites."[50] This meant that once mental integration was attained, there would need to be additional "distillations," alchemical processes entailing a reunion of mind with the body, and with the rest of nature. But mental purification had to come first. In this regard, Jung cited alchemist Gerard Dorn: One must "free the mind from the influence of the 'bodily appetites and the heart's affections'. . . . In order to bring about their subsequent reunion, the mind (*mens*) must be separated from the body . . . for only separated things can unite."[51]

As I see it, the secret of fluid passage from the *unio mentalis* to the subsequent stage of reunion with the freed mercurial body lies in the very structure of the Hermetic vessel. What would be discovered upon sealing that bottle in earnest? That it is a *vas pelicanicum,* a *uroboros.* Therefore, at the moment the bottle would be truly sealed, when Mercurius would be closed into it hermetically and no longer able to escape, one would find that the spirit would be outside the bottle as well, now as a beneficial agent of healing, of wholeness. By genuinely completing the vessel, by closing the body within the mind in an "airtight" fashion, the simple containment of body by mind would be overcome and the body set free in union with the mind. The body contained within the finished bottle, like the body of Aelia Laelia, would at once be *un*contained ("a body that has no tomb round it"). Nevertheless, this overview hardly does justice to the details and nuances of the stages of alchemical conjunction. I will have a lot more to say about that before I am finished.

In sum, the completed Hermetic vessel would be a structure that would contain itself, flow through itself, that would be "both content (mother liquid) and container."[52] The inside and outside of this remarkable bottle would be united paradoxically as a single side. And sealing the vessel hermetically in this manner—by moving it back into itself— would permit it to contain the backward movement of consciousness, the Proprioceptive passage in which the life energies earlier sent forth to support the ego's development would now be withdrawn. The perfected vessel would facilitate a Proprioception in which the Projections of basic ontological divisions—subject and object, space and time, life and death—would be consciously retracted in a retrograde circulation[D7] back into the hitherto unconscious Self. The vessel thus fashioned would be tantamount to the subtle body that the ancient Chinese hoped to create through their meditation exercises—the uroboric body of the "universal ego" or Self that would replace the ordinary physical body of the finite particular ego when that ego meets its demise.

But, alas, the alchemists of old were evidently unable to complete the hermetic closure of the vessel. The great stumbling block noted in the previous section was the tendency of magical thinking simply to *blur the distinction* between inside and outside or psyche and matter, rather than uniting opposites through the hermetic paradox the vessel demanded. The early alchemists indeed did not operate in the well-focused, sharply delineated space known to us today, but in a kind of waking dream space where boundaries tended to be fluid, including that between the practitioner and the substances with which he or she worked. So, if mercury, for example, were placed in a retort for distillation, the numinous psychic potency this substance had for the alchemist might "leak out" to fill him with a sense of awe and foreboding (like the youth encountering the powerful spirit Mercurius in the Grimms' fairy tale). Therefore, because alchemical "dream space" was considerably undifferentiated, the alchemist was given to experiencing within himself or herself the potent quality of the materials s/he worked with in a manner that could be quite alarming.

The alchemy of old had thus reached an impasse. Although the alchemist knew the importance of properly sealing the containment vessel, the magical thinking that prevailed thwarted this aim. What was needed was a new kind of space to surpass alchemical dream space, a space whose boundaries were defined well enough to permit objects to be sharply set off from one another, and to allow these objects in space to be unequivocally distinguished from the subject who observes and operates upon them. Eventually, the space would be brought into such razor-like focus that hermetic closure of the uroboric container would become possible. The space in question would then be paradoxically bounded so as to allow psyche and matter to be completely separated and at once completely fused! But the first order of business was to step beyond the oneiric space of the Middle Ages, and this meant that the uroboros would again need to be repressed in the interest of further differentiation.

THE RENAISSANCE, POST-RENAISSANCE CULTURE,
AND THE RETURN OF THE UROBOROS

In the previous chapter, I noted the profound transformation of consciousness that was wrought by the fourteenth-century revolution in European culture known as the Renaissance. Recognizing the crucial role of space in this, Gebser opens his introduction to the Renaissance by citing a 1336 letter of Francesco Petrarch, the discoverer of landscape. Petrarch describes his ascent of Mount Ventoux, an experience that "unsettles" him and produces "intense agitation": "Shaken by the unaccustomed wind and the wide, freely shifting vistas, I was immediately awestruck."[53] Petrarch's letter is written as a confession of an epiphany that transforms his life. He goes on to say that "what I experienced today will surely benefit myself as well as *many others*."[54] Gebser comments that the experience in question was nothing less than an "encounter with what was then *a new reality*. . . . The event that Petrarch describes in almost prophetic terms as 'certainly of benefit to himself and many others' inaugurates a new realistic, individualistic, and rational understanding of nature."[55] The trip to the summit of Mount Ventoux and subsequent inspection of the vast panorama that lay below is portrayed by Gebser as the first movement into a "free, open, and unenclosed space," an "elemental irruption of the third dimension and transformation of Euclidean plane surfaces."[56]

The space that Gebser alludes to is decidedly different from the earlier dream space of alchemy, with its fuzzy focus and fluid boundaries. Because the linkages formed through "sympathetic magic" involved directly apprehended resonances requiring no mediation, elements were joined in oneiric space through a kind of discontinuous leaping from one to the other wherein distinctions were blurred. But the new space of the Renaissance was a *continuum*. In such a context, there is no undifferentiated jumping across gaps but a point-to-point passage in which differences can be sharply delineated and details filled in. The points comprising the continuum are tightly juxtaposed to preclude any breaches or holes and this creates a well-sealed container. It was the unbroken continuity of the new space that permitted boundaries and distinctions to be drawn with unprecedented clarity, setting objects sharply off from their backgrounds and from each other.

The same transformative act that locked objects into space for clear-cut observation and measurement liberated the subject from space. According to post-Renaissance philosopher René Descartes, whereas an object is *res extensa* (a thing extended in space), the subject is *res cogitans* (a thinking thing), a purely mental being without extension in space and thus freely transcendent of it, not constrained by the physical laws that govern the objects. No longer bound to and caught up in the web of earthly ties, the post-Renaissance subject was "his own man." The newly self-possessed, autonomous individual, having extricated himself from worldly dependence, could now shape and mold the concrete world to his desires, (re)create it in his own image—activities that required him to cast the world before himself as an object in space (the word "object" comes from the Latin, *objicere*, "to cast before"). This brave new world ushered in by the Renaissance can be summed up as that of *object-in-space-before-subject.*[57]

However, while the revolution of the fourteenth century indeed paved the way for the dramatic transformation of space, and of the relationship between subject and object, several hundred years were required for the changes to crystallize and take hold. It is for this reason, I suggest, that the transition from the Middle Ages to the Renaissance did not simply bring the demise of old alchemy. Instead alchemy appeared to flourish with the work of Renaissance thinkers such as Paracelsus, Michael Maier, and Gerard Dorn. Yet, by the end of the seventeenth century, the influence of alchemy had in fact been undermined, and by the end of the eighteenth century, alchemy had been relegated to obscurity.

The century in which the downfall of alchemy began was the same century that brought the crystallization of the new space to apparent completion in the work of Galileo, Kepler, Descartes, and Newton. And this highly focused framework, constituting as it did a continuous context for measurement and analysis, was a key factor in the advances that followed. In principle, all bodies could now be precisely located, probed, and experimented upon, and submitted to mathematical scrutiny. This objectification of nature, this bringing of nature into focus as definitively representable in space and measurable by time, is what made the whole enterprise of modern science possible. From it came the development of more reliable research methods and fresh approaches to mathematics (e.g., the use of Cartesian coordinates, and of the infinitesimal calculus to analyze space-time motion), and these, in turn, gave rise to technological innovations that transformed our world.

Scientific progress continued in this way into the nineteenth century, reaching an unprecedented level of exactitude. Yet, ironically, this refinement of science exposed certain limitations of it that previously had gone unnoticed. A primary example of this was the work of Michelson and Morley on the physics of light. An accurate method for directly measuring the velocity of light had just been developed, allowing these researchers to conduct an experiment that would confirm the expectation that light was transmitted through space in a continuous way, like any other object (for more detail, see Rosen 2004, chapter 1). To the amazement of the scientific community, the expectation was upset, and this raised fundamental questions about space and time themselves. The whole post-Renaissance project had depended on the existence of a continuous medium in which the phenomena of nature could be probed and measured with complete certainty. Now—with the finding that conventional thinking about space and time could not deal with a phenomenon as basic as light—science's sense of certainty was seriously shaken.

Another finding of late-nineteenth-century physics posed an even greater challenge to the post-Renaissance tradition. The research of Max Planck led to the conclusion that the smooth continuity of nature given to ordinary observation breaks down when nature is observed on a microscopic scale. The inherent discontinuity of the physical world implies an inherent uncertainty about its fundamental processes. For, if we cannot continuously track the behavior of a particle through space and time no matter how we refine our measuring instruments, then, in principle, we cannot know its exact location or state of motion from moment to moment. Quantum mechanics suggests another way to look at the built-in uncertainty of nature. It is related to the idea that the very act of observing a particle unavoidably affects the particle observed. This interaction of observer and observed could be overlooked in the Newtonian era, since its influence was negligible at the relatively gross scales of measurement then employed. But developments in microscience brought physicists to the point where they were no longer able to deny these observer effects—effects that tended to subvert the empirical ideal of detached and objective observation. Modern physics therefore calls into question what are perhaps the two most basic tenets of post-Renaissance science: the continuity of space and time, and the separation of the observer and observed, of subject and object. In short, what we find at the heart of contemporary physics is a fundamental challenge to the Renaissance formula of object-in-space-before-subject.[58]

The disturbance of Renaissance order was not limited to science. In the 1880s—the same decade that physicists Michelson and Morley were calling classical space into question—space was also being disrupted in the field of art. Édouard Manet's *A Bar at the Folies-Bergère* (fig. 2.4) is my case in point.

Figure 2.4 Édouard Manet's *A Bar at the Folies-Bergère* (1882).

Art historians Paul Vitz and Arnold Glimcher say that Manet's *Bar* is an example of *fractured space*: "two or more discrete lines of view [are] present at the same time in a given portrayal of space: these separate but simultaneous views break or fracture what was once (seen as) homogenous."[59] In *Bar*, the left two thirds of the picture give a view of the woman from a frontal perspective, whereas the mirror image of the woman's back and the top-hatted gentleman shown in the right third of the painting suggest an angle of vision displaced to the left of center. Vitz and Glimcher note that this juxtaposition of opposing perspectives leaves the viewer in an ambiguous state. The undercurrent of visual tension and uncertainty that accompanies the viewing of *Bar* arises from the fact that while your attention is occupied by one perspective, at the same time you subliminally feel the pull of its overlapping opposite. The experience here is quite different from what is encountered in Renaissance art, where a single perspective confirms the existence of a unitary visual space.

There are many other examples of nineteenth-century breaches in Renaissance order. Not only was confidence shaken in the old sense of spatial continuity, but in the continuity of time as well. Berger and Mohr describe the advent of photography as bringing a "shock of discontinui-

ty," since the instantaneous flash that produces the photograph "arrests
the flow of time" in a way that drawings and paintings (which are produced over a duration) do not.[60] Here the continuous passage from the
past into the future that had characterized earlier temporal experience
was ruptured by a decontextualizing perception of instantaneous leaping.

Revolution also shook the field of mathematics in the nineteenth century. Post-Renaissance science had relied on the postulates about points
and straight lines that formed the basis of Euclidean geometry. But with
the work of mathematicians such as Bolyai, Lobachevsky, and Gauss,
questions were now raised about Euclid's parallel postulate (which states
that one and only one parallel line can be drawn through a point external
to a given line). This resulted in the development of non-Euclidean alternatives, nonclassical spaces characterized by various kinds of curvature.
Importantly, there was no intrinsic basis for determining which alternative reflected what was "objectively real."

Before the nineteenth century ended, the mathematical assault on the
post-Renaissance tradition of objective realism was driven to the hilt.
Though the non-Euclidean geometries indeed had departed from space
as classically intuited, they had preserved its most fundamental feature,
its *continuity*. However much space might be curved, warped, or distorted by the new geometries, it was not ruptured, not broken apart.
Yet that too came to pass. Space was torn asunder by grossly nonintui-
tive structures such as the "Cantor set," which brought the decimation
of traditional perspective to its logical conclusion: while the points of
perspectival space are bound together in a unitive framework that permits only one frame of reference to be adopted at a time, each point of
"Cantor space" is, in effect, a law unto itself, so that an infinite number
of differing perspectives could be simultaneously assumed! Structures
such as these, which fracture Euclidean space and fly in the face of post-
Renaissance intuition, have been referred to by some mathematicians as
a "Gallery of Monsters" (see Mandelbrot, who has attempted to describe
these "monstrosities" with his fractal geometry).[61]

My final example of the nineteenth-century challenge to earlier
tradition comes from the field of philosophy. The twin pillars of post-
Renaissance philosophy are empiricism and rationalism. The former
seeks objective knowledge through experience, especially systematic
observation via the senses. For the latter, knowledge is gained through
the power of reason, with particular emphasis on intellectual and deductive operations. At bottom, both approaches adhere to the old formula
of object-in-space-before-subject. Post-Renaissance sensory perception

primarily entails perceptions of objects and objectified events arrayed in the spatial continuum, these empirical observations being carried out by detached subjects. Deductive rational operations can be described in a basically similar way, except that the "objects" of knowledge are manipulated in a more abstract mental space or cognitive continuum and the rationalist subject is even more removed from concrete reality than his empiricist counterpart.

Rationalism and empiricism alike were lampooned by nineteenth-century existentialist writers such as Kierkegaard, Nietzsche, and Dostoyevsky. The case of Søren Kierkegaard is particularly interesting in view of our discussion of physics, since this denizen of Copenhagen had a significant influence on another prominent inhabitant of the city: Niels Bohr, a prime architect of quantum mechanics.[62] I find it intriguing that the key features of quantum physics—discontinuity, uncertainty, and subject-object interrelatedness—were evidenced decades before the inception of that new science in the mid-nineteenth-century philosophy of Kierkegaard. The Danish thinker passionately dedicated himself to attacking the dispassionate posture of classical science, and of post-Renaissance culture in general. Kierkegaard encouraged intimate relations with others and the embrace of uncertainty and even paradox, and his own discontinuous leap of faith foreshadowed the quantum leaping of subatomic particles.[63]

I would like to suggest that the revolutions of the nineteenth century signify the return of the uroboros. That the long-repressed symbol of paradoxical wholeness should reassert itself at this particular time is neither arbitrary nor especially mysterious. It is as Neumann said: when the process of individuation reaches sufficient maturity, the symbol of the uroboros will reappear. That seems to be what happened in the nineteenth century. The post-Renaissance order of object-in-space-before-subject had served the purpose of further differentiating and refining the ego, as medieval alchemy had required. The precision that was brought through nineteenth-century advances in science and related fields reflected the development of an ego now mature enough to recognize the limitations of the classical system it had relied on so completely. Thus it was that science's new and highly refined tools of measurement opened to question the ideal of the spatial continuum and with it, the apparent certainty this framework had brought for the detached and objectifying ego. Questioning post-Renaissance order was unnerving indeed, for it had been the mainstay of human identity for several hundred years. This was the order that had been projected so as to extricate the individual from the protean dream space that previously mingled subject and object in an indiscriminate way; it was the order through which the uroboros had been eclipsed in the interest of further individuation. The glimpse of the uroboros now gotten through cracks in the post-Renaissance façade was no doubt threatening to behold.[D8]

What was the prevailing reaction to the nineteenth-century intimation of the uroboros? Clearly the individuation process had advanced to the point of being able to register the presence of the uroboros, at least indirectly or semiconsciously. But could this archaic symbol of the transpermeation of subject and object, of life and death themselves, be *accepted*? Were we finally prepared to switch gears, to counteract the forward thrust of post-Renaissance Projection and engage in the Proprioceptive movement backward into the Self? Not surprisingly, the answer is no. Too much had been invested in the ideal of the freestanding, self-possessed, self-certain ego for said ego to accept its demise with equanimity. So the challenge of the resurgent uroboros was met across Western culture by the massive act of denial that has come to be known as *modernism*.

Most essentially, modernism seeks to suppress the uroboros and preserve the post-Renaissance order by moving to higher levels of abstraction. Therefore, when the classical order of continuous space and time was cast into doubt by the Michelson-Morley experiment, Einstein transformed physics by proposing a more abstract space-time continuum, yet one that still served as a field of objectification for the detached subject. Planck's microphysical discontinuity was similarly smoothed over by a pseudo-continuous abstraction of space (the Hilbert space), and the "shock of discontinuity" of the photograph was eased by the simulated continuity of the cinema. In art, the ambiguities of Manet and Cézanne were surpassed by the abstract certainties of cubism, and, in philosophy, the existential angst of Kierkegaard and Nietzsche gave way to the "objective" scientific philosophies of the twentieth century. These and other examples of the transition to modernism are examined in detail in *Dimensions of Apeiron*.[64] There I attempt to show that the aim of modernism in all cases is to maintain the post-Renaissance Projection of object-in-space-before-subject.

Also explored in *Apeiron* is the *collapse* of the modernist project: its failure in science and its postmodern disintegration in the media, the arts, and culture at large. Modernism's downfall is further investigated in *The Self-Evolving Cosmos*, where I focus on the current impasse in modern physics stemming from its continuing penchant for *objectifying* nature in the face of phenomena that no longer lend themselves to that.[65] Whereas my earlier works look primarily to phenomenological philosophy (à la Maurice Merleau-Ponty and Martin Heidegger) as a constructive alternative to the objectivist philosophy undergirding modernism, what I want to emphasize in the present context is that our 150-year-old crisis of individuation can be addressed beneficially by picking up the threads of alchemy (the connection between phenomenology and alchemy is intimated in *Apeiron* and in Schwartz-Salant's *The Black Nightgown*).[66] I venture to say that the nineteenth-century upsurgence of the uroboros was an invitation back to alchemy that could be extended at that time because science had completed its work of surmounting the deficiencies of old alchemy. A century and a half later, having witnessed the fate of modernism in refusing the invitation, we would do well to finally accept it. Indeed, a few courageous individuals have already heeded the call.

Among the most prominent twentieth-century advocates of alchemy was psychologist Carl Jung. His work and personal life were profoundly influenced by the hermetic science and he devoted several full volumes to the subject.[67] Moreover, Jung was convinced that the efficacy of future science depended on adopting a revitalized alchemical approach. According to his personal secretary Aniela Jaffé, Jung called for "the construction of a new, unitary world-model [in which] spirit and matter are no longer opposites. . . . This new world-model is a reconstruction of the old, intuitive vision of the alchemists."[68] Jung anticipated the demise of conventional science and reemergence of alchemy with the following words:

> The moment when physics touches on the "untrodden, untreadable regions," and when psychology has at the same time to admit that there are other forms of psychic life besides the acquisitions of personal consciousness—in other words, when psychology too touches on an impenetrable darkness—then the intermediate realm of subtle bodies comes to life again, and the physical and psychic are once more blended in an indissoluble unity. We have come very near to this turning-point today.[69]

Jungian analyst Nathan Schwartz-Salant adds his own insights on the history and current relevance of alchemy, and sheds new light on the nature of alchemical relatedness. In *The Mystery of Human Relationship*, he explores the great promise of alchemy but begins by acknowledging the earlier deficiencies that led to its repression at the hands of rationalist science: "alchemy was totally unsuited to understanding nature in causal terms, as postulated in the great scientific advances of the seventeenth and eighteenth centuries."[70] Schwartz-Salant goes on to speak of the "wild misuse of subjectivity and fantasy that the alchemical approach can manifest," and he notes that "the alchemists' understanding and transformation of matter was . . . far inferior to what has been accomplished by modern science. . . . The alchemical approach to the world gave priority to a background sense of oneness which kept it from ever successfully separating from and adequately evaluating its most important tool: the imagination." Schwartz-Salant is therefore led to conclude that

> the ideas and practices of [old] alchemy . . . had to retreat to allow the individual ego to develop, an ego that could believe it was separate from other people and the world. . . . During the time of the emergence of alchemy in the [Middle Ages and the] Renaissance, ego-consciousness had barely developed. But without the careful discrimination of the ego, [one] readily regresses into a hopeless muddle of fusion that blurs any subject-object differentiation. The mind of the Renaissance and before, was characterized by an immersion in images and by a lack of critical reflection.[71]

So the transformations of science and human intellect that came after the Renaissance clearly were needed. "Descartes' separation of mind and body as two qualitatively different entities" was "radical but necessary. This great achievement in consciousness—an objectivity about nature and the development of self-awareness—which began in the seventeenth century made possible the modern scientific approach."[72]

Nevertheless, today something different is required:

> Today, we must recognize the shadow side of the great development of ego consciousness, namely the creation of defenses that allow too much separation of the ego from the unconscious and from the emotions of the body. Alchemical thinking holds out a way of returning to wholeness without abandoning separation and distinctness of process. In a way alchemy's time has come. Perhaps we can now return to those mysterious realms or "third areas" that are neither physical nor psychic, domains whose existence must be recognized if we are to reconnect split orders of reality such as mind and body. I believe that such "third areas," a

major concern of alchemy but left behind by [classical] scientific thinking, will have to be reintroduced.[73]

I expect Schwartz-Salant would agree that the problem of revitalizing alchemy to meet the challenges of the contemporary world is largely one of *containment*. It was old alchemy's inability to contain the "Spirit Mercurius" that led to its repression after the Renaissance. But the Spirit was "out of the bottle" again once science reached fruition in the nineteenth century—only to be tenuously reimprisoned by modernism. Now, with modernism's collapse, uroboric Mercurius has been unleashed once more. The difficulty of containing the uroboros is implicit in Schwartz-Salant's description of the nature of the "third area" that would fuse psyche and matter while at the same time keeping them separate:

> [The] "third area" . . . cannot be experienced or understood through the spatial notion of insides and outsides. . . . [It] has its own peculiar objectivity: a subjective-objective quality. . . . [Here there is] a paradoxical sense of space in which one is both inside and outside, an observer but also contained within the space itself. . . . We must move beyond the notion of life as consisting of outer and inner experiences and enter a kind of "intermediate realm" that our culture has long lost sight of and in which the major portion of [alchemical] transformation occurs.[74]

The neo-alchemical employment of the "third area" to bring into harmony the stark ontological oppositions of post-Renaissance culture (subject and object, mind and body, spirit and matter) would entail nothing less than the effective containment of Mercurius.

In early alchemy, the shape-shifting Mercurius is not infrequently portrayed as a hermaphrodite. Symbolically, this figure constituted a total merger of masculine spirit or mind and feminine matter (mater, mother) or body. Mercurius could not be contained when these two aspects of its nature were blended so indiscriminately that the mind of the alchemist was always susceptible to being flooded by the chthonic potencies of the uroboric mother. That was why seventeenth-century rationalism was needed to drive a wedge between "father-spirit" and "mother-matter" (as von Franz put it).[75] But what was intimated in the nineteenth century, and again with the twentieth-century failure of modernism, is that rationalism's containment of Mercurius was actually incomplete. Enter here the contemporary alchemist. The "third area" of which Schwartz-Salant speaks offers a paradoxical, "inside-out" kind of containment in which mind and matter are hermetically sealed off from each other at the very same time that they fully interpenetrate. If this sounds familiar it is because we were dealing with basically the same issue in the previous section. The structure of the "third area" is none other than that of alchemy's Hermetic vessel. Alternatively described, it may be associated with the subtle body.

We saw above that Jung anticipated the revival of alchemy as a time when "the intermediate realm of subtle bodies comes to life again, and the physical and psychic are once more blended in an indissoluble unity." For his part, Schwartz-Salant refers to the "merger of the inner reality of the alchemist and the outer reality of the matter to be transformed" as occurring in "a space that alchemists called the 'subtle body,' a strange area that is neither material nor spiritual, but mediating between them."[76] Our own first encounter with the subtle body was near the beginning of the present chapter where we explored its relationship to the theme of death and resurrection.

The ultimate goal of old alchemy was the overcoming of death, and it was believed that this could be achieved by fashioning a subtle body, a "body of a spiritual kind."[77] Death was the most formidable chthonic power, the underworld uroboric potency that no ordinary being could contain. For, such a being is limited by its deep identification with a finite body, and this assures its eventual destruction. What was needed was an *infinite* body, a uroboric body.[D9, D10] That is to say, one had to *become* the mercurial uroboros in an embodied way in order to withstand the ravages of death. In surpassing the gross body to create its subtle counterpart, one was to shift one's identity from the body experienced as something merely physical, a circumscribed object, to the *psycho*-physical, *sub*-objective body of the uroboros itself. Here there would surely have to be a death: the death of the being that is wholly identified with its finite body, the death of the ego.

In this perilous quest, one dare not behold the uroboros as if it were just an object appearing before one. One dare not directly set one's eyes on the face of this Medusa—she with hair of twining serpents—lest one be turned to stone. What was in fact required in drawing close to the uroboros, in taking her body as one's own, was a movement *backward* into that infinite body, the Proprioceptive movement described above. It could conceivably be said that the myth of Medusa gives an indication of this in the idea that the frightful Gorgon could be safely observed by turning one's back on her and viewing her in a mirror. However, in the Greek myth, the mirror actually was used by the hero Perseus simply to avoid a stony fate and slay Medusa, an act symbolic of repressing the uroboros in the interest of the ascendant patriarchal ego. Given the alchemist's aim of fashioning the subtle body, his attitude toward the Medusa had to be more subtle. Though the alchemist of old might not have been able to fully achieve it, what was called for was viewing the Medusa's reflection while moving backward into the embrace of this uroboric figure. By so gazing into the glass, the alchemist presumably would glimpse his or her own reflection merging with the uroboric one. In this way, the stony ego death shunned by Perseus would be welcomed, but not as a mere extinction of being. Instead this death would bring a resurrection in which the uroboros's base stone, its *prima materia*, would be transformed into the Philosopher's Stone.

In *Mysterium Coniunctionis*, Jung in fact portrays the alchemical opus as a series of conjunctions culminating in the realization of the Philosopher's Stone. The final stage "of the opus alchymicum was indubitably the production of the lapis."[78] The Stone, says Jung, is a "symbolic prefiguration of the [S]elf," and to prepare for the ultimate Self-consummation, one must realize oneself as "'living stone,'" a "'stone that hath spirit.'"[79] We also learn from Jung that the Philosopher's Stone is linked to the *unus mundus*. Accordingly, "Dorn sees the . . . highest degree of conjunction in a union or relationship of the adept . . . with the *unus mundus*."[80] Indeed, the *unus mundus* (the "potential matter of the first day of creation") corresponds to the "prima materia," which, elsewhere, Jung correlates with the lapis.[81] Linking the uroboric Stone to the *unus mundus* in this way confirms the Stone's intimate relationship to *death*, for we have established that the *unus mundus* is the "realm of the dead."[82] Although Jung himself does not explicitly make this connection as far as I know, his colleague and collaborator Marie-Louise von Franz does just that in her book, *Number and Time*, as we discussed in the first section of this chapter. Thus, the creation of the subtle body ultimately entails a backward passage into the underworld sphere of the uroboros-as-Philosopher's Stone. The fashioning of this infinite body is the key to the ego death and rebirth that lies at the heart of alchemy.

Yet we have seen that, before the backward movement can be brought to completion, it is first necessary to move all the way *forward*. The uroboric body of paradox that fuses opposites without confusing them could not be contained were it not for Renaissance and post-Renaissance revolutions in science, mathematics, and philosophy that have contributed to the sharpening of ego consciousness and have brought us to the refined stage of individuation we have reached today. But still more needs to be done by way of honing our discriminatory powers vis-à-vis the subtle body if it is to serve effectively in containing the paradox of death and resurrection. To be sure, in the course of completing this further refinement, the gearing of individuation will have to shift from forward into reverse, from the posture of Projecting object-in-space-before-subject to its Proprioception.

DREAM JOURNAL

D1

I'm wondering just how much of the present text remains driven by the ulterior motives of my own insular ego. Is the present book not still part of my lifelong effort to call attention to myself by performing impressive feats of creative intellect, even attempting to achieve the "impossible" in the endeavor? My dream of January 26, 2010, reflects this deep-seated tendency in myself—and ends by calling it into question:

During a gathering at a friend's house, I am displaying a tremendous breadth of metaphysical knowledge, especially about added dimensions. One young fellow who has heard me speak is extremely impressed by what I've said. I talk to him about other dimensions, and answer questions that he never thought could be answered. At one point he asks me the following question, not expecting I could possibly know the answer to it, given how difficult and abstruse the question is: "What is the true nature of the reality existing on the far side of objects, the side we can't see?" I say, "You're talking about the Laws of Perspective. The far side of objects involves the fourth dimension."

The young man is so impressed by me that he wants to know if it would be okay for him to call me. Asking for my telephone number, he tears off a piece of paper and I write the number down for him. My feeling is that I don't mind if he calls me, but I wind up thinking: What does it really matter? What am I getting out of it? I don't need it, and don't really want to start another relationship of this kind with somebody, based on how thoroughly impressed they are with my mental prowess.

The theme of "doing the impossible," approaching the infinite, is echoed in an older dream . . .

D2

Dream of October 29, 2007:

I am enrolled in some kind of academic course, and am reading the instructor's response to a paper I've written. The teacher indicates that my work reaches the quality of "Master Carnobis." Apparently this means that the caliber of the work far surpasses what is expected for a course at that level. I start rereading the paper and realize that—yes—what I do here is achieve transcendent heights, approach the infinite, fly. I now see myself swinging high and free, in great circular arcs, as if on a trapeze.

This dream suggests the dizzying ecstasy of something transcendent, and brings to mind a much older dream . . .

D3

Dream of May 3, 1991:

Scene 1
A woman says to me very affectionately, "Come lie down with me, sleep with me." I don't want to do it, I really don't. I feel guilty about this. The woman is dying. She is my mother.

Scene 2
I am in my parents' bedroom and they are sleeping. I am trying to do something that would impress them: mount a ladder. I want the ladder to stand straight up. The exercise is closely associated with doing yoga, particularly the pose of standing on one foot with one arm stretched upward toward the ceiling. In the dream, I have to mount the ladder in my parents' bedroom. There are two foam rubber earplugs that I am supposed to wear while I do this (much like the earplugs I've actually used in doing yoga), but they keep popping out.

Willpower is an element in the exercise: I have to keep the ladder erect, get it to stand straight up—but it keeps lapsing back. At one point, I have to do this from an impossible position, one in which the ladder is slanted on an acute angle with the floor and I am clinging to a rung with my back hanging down.

But I actually do get the ladder to straighten out. This is a paranormal event, a transcendental accomplishment, and I start to feel tremendous

energy surging up in me, as if I am going to levitate. Then I stop myself.
Yet I do want to impress my sleeping parents—especially my mother. I
want her to see me levitate; if she sees this, she'll be proud of me.

Scene 3

I am lying on my back sleeping deeply, and my mother comes into my
room and is approaching my bed. She is in a stage of half dress, wearing a
long black slip and matching black top, as if she is getting dressed up to go
somewhere. As she draws near to me, I panic and attempt to pull myself
out of the dream, to wake myself up by moaning.

I believe this dream uniquely maps my process of individuation. In scene
1, I'm under my mother's sway, feeling her strong gravitational pull. She
wants me to lie with her, to sleep with her, and I feebly resist.

The next scene has me projecting myself out of the feminine orbit
into an upright posture, a phallic thrust intended to impress mother with
my power to do the impossible. Through force of will, I am to levitate, to
stay erect in defiance of the laws of gravity. After having great difficulty
maintaining my "transcendental erection," I am on the verge of success.
Through this colossal feat, I will make my mother proud of me. And I
will win motherly approval on *my own terms*—not by succumbing to her
influence but by breaking free, projecting myself outward and upward,
projecting my ego into the infinite.

There is obviously an aspect of ego inflation here, the fantasy of soar-
ing beyond the earthly sphere of mother's body to attain supernal bliss.
But I suspect there is more to this dream. For just when I am about to
levitate, I stop myself. Why do I do that? Do I not want my mother to
witness my paranormal accomplishment? Is that not the whole purpose
of my efforts to defy gravity? And yet—feeling a great surge of energy
well up in me in advance of levitating—I shut the process down. Why?
I propose it is because the energy surge short-circuits my egoic fantasy
of realizing the infinite and confronts me with its electrifying actuality.

When I had the dream in 1991, I didn't recognize its connection to an
experience I had twenty-three years earlier. On a night in 1968, I turned
off the light beside my bed and drifted into a hypnagogic vision of moving on a sled across a large field of snow. Gliding soundlessly through the
soft whiteness, a sense of serenity enveloped me. The sled then began to
rise into the air, to levitate. As this happened, I was seized by a feeling of
exhilaration that quickly built into the sense that every cell in my body
was about to explode in unbearable ecstasy! The painful bliss became

so excruciating that, by a conscious effort of will, I shut the experience down—just as I shut down the levitation experience in the dream I had many years later.

I have come to understand my hypnagogic vision of 1968 as a spontaneous *kundalini* encounter—the encounter that started me on my long journey to the writing of this book. In chapter 7, kundalini energy plays a central role, and we see its intimate relationship to the uroboros, for the kundalini phenomenon is traditionally personified as a coiled serpent that swallows its own tail. Bearing in mind Erich Neumann's association of the uroboros with the feminine archetype of the Great Mother, what conclusions am I led to in my attempt to understand the dream?

The dream can be related to Neumann's synoptic take on the individuation process: "The same uroboric symbolism that stands at the beginning, before ego development starts, reappears at the end, when ego development is replaced by the development of the self, or individuation." The dream begins in the presence of the uroboric Mother. I then seek to project myself above and beyond the Motherly realm. And yet, just when I seem on the threshold of achieving my goal of phallic freedom, the uroboros makes her presence felt again, heralded by an upsurge of kundalini energy. I am not prepared for this. I am not yet ready to drop the egoic ideal of transcendental fulfillment for the actuality of the embodied Self. So I stop the process. Unable to accept the demise of my insular ego and to embrace the uroboros, in the final scene I retreat into deep sleep, only to react with panic when she approaches me. What Neumann says of individuation in general applies to the dream: the same uroboric symbolism that stands at the beginning reappears at the end. But I must *embrace* the uroboric Mother if individuation is to proceed, and this is something I have not yet fully done.

D4

The theme of death, resurrection, and wholeness suggests itself to me in my dream of November 12, 2005:

After teaching my classes at the college, I am on my way home. I'd been told I might experience some delays, but it is smooth sailing as I drive eastward toward the Verrazano Bridge. Then, nearing the bridge, I encounter an obstacle. There is a huge excavation site with even layers of fine brown soil filling the bottom. This massive dirt hole (I originally wrote "whole"!) is shaped in a neat square, and is obviously under preparation for the construction of something new. It is quite impossible for me to bypass the excavation to get to the bridge. (In recording the dream, I realize what the bed of brown earth reminds me of: a grave. It is what a gravesite looks like before a body is lowered into it.)

At this point in the dream, there is a vague sense of being accompanied by my wife, Marlene. Unable to depart Staten Island, we need a place to stay, so we go up a mountainside to a motel. Once there, I run into a colleague of mine, an older philosopher whose work is widely known. Speaking casually and in his typically self-possessed fashion, he comments that he quite likes this motel, and that it is a lot better than it used to be.

At a certain point, the philosopher and I hug each other. When this happens, it seems as if our bodies merge. In the prolonged embrace, it is as if we are fused into a single being, so that the philosopher's body becomes an integral part of my own. What a feeling of ecstasy this brings! He is present. His strength is there for me. And I sense with certainty that it always will be.

Upon awakening from the dream and for much of the morning, I feel the afterglow of that embrace. It fills me with joy every time I think of it. In retrospect, it seems clear that the obstacle I encounter earlier in the dream is my death: a bed of brown earth has been prepared—a burial ground. But this is also a *construction site*, and the ground has been cleared for the creation of "something new." Is this brought to fruition on top of that mountain? Could this be a dream of death and rebirth providing a glimpse of the *unus mundus*?

D5

My dream of October 27, 2004, focuses on the laboratory preparation of a strange "hybrid phenomenon":

I need to pick up an order on a university campus and I'm in a big hurry. I'm looking for shortcuts, trying to use my guile to get the job done quickly. I attempt to convince a cop to let me park my car in a restricted area convenient to my errand, but he's indifferent to my effort to charm him.

After finding proper parking, I locate the right university building and arrive at a laboratory where my order is being filled. Technicians are preparing a peculiar-looking creature submerged in a large tank of water. The being seems oddly out of focus to me. It is composed of a jumble of varying shapes and colors and has an unformed, ragged appearance. While some of its features are box-like and geometric, as if produced by human design, others seem sloppily organic, like the slimy swirling tentacles of an octopus.

I ask one of the technicians how long it will take before the work is finished. He says it will be at least two hours, and I'm taken aback. Does that mean I'm going to be wasting the whole evening on this? Feeling put out, I try to get him to accommodate me by speeding things up, but to no avail. I ask him what I'm going to do here for the next two hours and he suggests I leave and come back. Grudgingly, I resign myself to the hassle.

After recording the dream, I associated the hybrid creature being prepared in the laboratory tank with alchemy's *subtle body*. Can such work be done in a hurry? Can it be achieved by cutting corners, taking shortcuts? In fact, it could plausibly be said that this whole book deals with fashioning the subtle body—a process that surely can't be rushed!

D6

Perhaps I had my own dream of "Aelia Laelia" on October 16, 2003:

A black woman is giving a magnificent performance. She voices her inner emotions in a highly evocative way, and then—as if standing outside of herself and observing—she describes her reaction to those feelings. "She's turning herself inside out," I think. I am astounded at what she is doing and feel great admiration for her. Her performance is electric. Its vividness and energy have a numinous quality.

D7

My dream of July 3, 2002, seems to be telling me that, in the end, there is no *safe* way to enact the retrograde circulation that makes the unconscious conscious:

I am trying to escape from a woman who frightens me. The idea is to find my way home, to safety, by a kind of circular movement that runs counter to the usual routes. I have to work my way back in an unexpected fashion if I don't want to be discovered. This involves moving through darkened streets that are off the beaten track. These side streets or back streets are familiar to me.

*Is there a promise of safety in these streets? The path to safety is off
the beaten track, through the familiar, the unobtrusive, the inconspicuous.
Apparently, I will not be noticed if I slide inconspicuously into these back
spaces, working my way home in this manner. To avoid exposure to the
force of the woman I so greatly fear, I have to play it safe.*

*In the dream's finale, I'm climbing a circular or spiral staircase, moving
counterclockwise. Suddenly, I sense that, coming down from above and
moving clockwise, is a woman holding a baby or child. I am so panicked by
this that I wake up moaning.*

The dream may speak to my ambivalence about my work on the unconscious: while it feels quite safe to think and write about making the
unconscious conscious, actually *doing* it terrifies me!

D8

Dream of March 6, 2004:

*I and others are warding off violent attacks in the dark. We are engaged
in a desperate battle to save our lives. Then the lights go on and I can
see many people who, in one way or another, are defending themselves
against attacking snakes. There are several snakes that are being held off
at arm's length, or foiled in other ways.*

D9

The uroboros can be expressed as a symbol of infinity, or, rotating the
∞ sign, as a figure-8 configuration (see fig. 7.4). And the figure 8 evokes
a relevant dream. In her book *On Dreams and Death*, Marie-Louise von
Franz reports a dream she had several years after her father's death, in
which the number 8 played a prominent role:

*I was with my sister and we both wanted to take Tram No. 8 at a certain
place in Zurich, to go to the center of town. We leaped onto the tram and
discovered too late that it was going in the opposite direction. I said to my
sister, "If one of us had done this it would be just a mistake, but since both
of us have done it, then there must be a meaning in it. Let's watch out for
what it may lead to." ... At the next tram stop we got off and there a taxi
drove up near us and out of it—came my father! I knew it was his ghost.
When I started to greet him he made a sign not to come too near to him*

and then walked away to the house where he had lived. I called after him,
"We don't live there any more." But he shook his head and murmured,
"That doesn't matter to me now."

Von Franz identifies "Tram No. 8" as a key element of the dream: "In number symbolism, eight represents timelessness and eternity. . . . In alchemy eight is the number of completion."[83] And it is clear that the retrograde movement described in the dream (the tram was "going in the opposite direction") involves a passage into the realm of the dead. Here a dream of my own comes to mind . . .

D10

Dream of May 7, 2005:

I'm away on a trip somewhere with a group of people. For some reason, I must leave early, travel back. I really don't want to cut my stay short and go off alone at night, but I can't stay with them. I must be on the train back by 8 o'clock that evening. It would be very easy to miss this train, and I'm worried about it. I've got to go somewhere, go back, so I have to be careful not to miss the train.

I'm supposed to eat a meal before I go, but there doesn't seem to be enough food around. Some people want to help me. There's a feeling of feminine support. Yet my situation still feels difficult: I can't stay. I do have to pack up my things and make the 8 o'clock train back, but I've got to eat before I do this. The problem is that I can't eat. There's just not enough food.

This dream saddens me, makes me weep. I deeply regret having to break away from the people I came with, abort my trip, and leave alone in darkness. There's a sense of deprivation, of being exiled from the light. But it can't be avoided. Does leaving alone in darkness mean *dying*?

Why must I make the 8 o'clock train in particular? Why would it be "so easy to miss"? Why am I so anxious to avoid that? And why is there not enough food to bolster me for the journey back?

To repeat what von Franz said after describing her dream of "Tram No. 8" and associating it with death: "eight represents timelessness and eternity. . . . In alchemy eight is the number of completion." Though death is inescapable, not every death entails alchemical completion. In fact, an alchemical death, one involving a realization of the infinite, is indeed very easy to miss. So perhaps the "8 o'clock train" signifies for me the infinite Self to which I want to return, though I may need special nourishment for the task. Perhaps consciously realizing the Self by fleshing out the subtle body requires that I feed myself in the fashion of the uroboros: I must nourish myself *on* myself. That is, I must "swallow" myself in the sense of moving backward into myself with awareness, until the Self per se awakens in me to Proprioceptively withdraw the Projection of the finite being that I am. This is how I make the "8 o'clock train": I die consciously as a finite being and am reborn as the Self.

The dream then seems to echo the central theme of this whole book.

CHAPTER THREE

Alchemy and Topology

This is a tomb that has no body in it
This is a body that has no tomb round it
But body and tomb are the same

These words from the enigma of Bologna darkly intimate the underlying design of the subtle body.[1] Presently, we will attempt to bring the inside-out structure of our infinite body into sharper focus, distill it with greater precision.

THE NECKER CUBE

To begin, let us consider the added detail provided in Schwartz-Salant's description of the realm in which the work of alchemy took place:

> It is an "in between" world of "relations," occurring in a space that is not Cartesian, and instead is characterized by a paradoxical relationship in which "outer" and "inner" are alternatingly both distinct and the same. . . . The paradoxical geometry of this space [is] known as the subtle body . . . an "intermediate" realm between matter and psyche.[2]

By way of further specifying the "paradoxical geometry" of the subtle body, we turn to modern psychology's study of perception, where we find a curious figure known as the Necker cube (fig. 3.1b).[3] To appreciate the remarkable properties of this structure, consider first the Cartesian principle of opposition, as expressed through visual perspective in figure 3.1a.

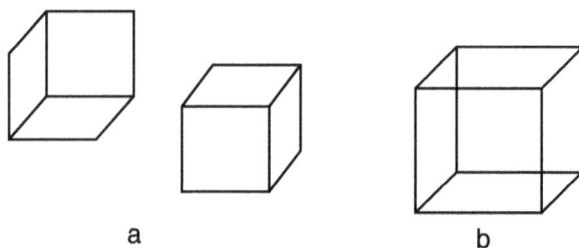

a   b

Figure 3.1 Opposing perspectives (a) and Necker cube (b).

If you were initially viewing a solid cube from the angle shown in the left-hand member of figure 3.1a, you would obtain the point of view of the right-hand member by (1) moving 180° around the cube to the opposite side, and (2) moving above the cube, since the left-hand perspective gives the view from below. The faces of the symbolized solid that are visible from the right-hand perspective are precisely those which were concealed from the left-hand point of view, and vice versa. In our ordinary experience with perspective, it is of course impossible to view simultaneously both the near and far side of an object, or its inside and outside; all the faces of the cube cannot be apprehended in the same glance. Opposing faces are closed to each other.

The ordinary mode of perception is the Cartesian one. Here we perceive objects and events as extended in the world outside us, but have no immediate access to the inner, subjective ground of our perceptions: we cannot see our own act of seeing, touch our own touching. What figure 3.1a illustrates is that this underlying opposition between the subjective seat of perception "in here" and the objective realm "out there" is reflected in the external objects themselves, in the diametrical opposition we normally encounter between their concealed and exposed surfaces. Opposing sides of objects cannot be viewed at once.

Turning now to figure 3.1b, you can see that both of the perspectives shown in 3.1a are encompassed in the body of the Necker cube. This creates visual ambiguity. You may be perceiving the cube from the point of view in which it seems to be hovering above your line of vision when suddenly a spontaneous shift occurs and you see it as if it lay below. Two disparate perspectives certainly are experienced in the course of gazing at the cube and this disparity reflects the continuing distinction between opposing sides. But the cube's reversing perspectives overlap one another in space, are internally related, completely interdependent (think of what would happen to one perspective if the other were erased!).

To approximate more closely the paradoxical geometry of the subtle body wherein "'outer' and 'inner' are alternatingly both distinct and the same," let us attempt to go a step further in our perception of the cube. Ordinarily, our glance is limited to merely oscillating from one perspective of the cube to the other. But we can actually break this visual habit and view both perspectives at once.

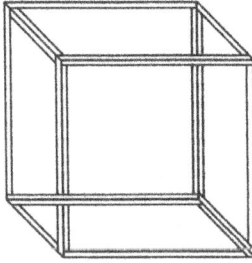

Figure 3.2 Necker cube with volume.

In figure 3.2, I've added some volume to the Necker cube, fleshed it out a bit. This modification should make it easier for you to see what I am talking about. When the cube's perspectives are integrated as I am suggesting, there is an uncanny sense of self-penetration; the cube appears to do the impossible, to go *through* itself. Here the division of sides is surmounted in the creation of an experiential structure whose opposing perspectives are simultaneously given.

But the word *simultaneous* may not exactly fit. I am proposing that we can apprehend the cube in such a way that its differing viewpoints overlap in time as well as in space. Yet what we actually experience when this happens is not simultaneity in the ordinary sense of static juxtaposition. We do not encounter opposing perspectives with the same immediacy as figures appearing side by side in Cartesian space, figures that coexist in an instant of time simply common to them (as, for example, the letters of the words on this page). And yet, there is indeed a temporal coincidence in the integrative way of viewing the cube, for perspectives are not related in simple succession (first one, then the other) any more than in spatial simultaneity. If opposing faces are not immediately co-present, neither do they disclose themselves merely seriatim, in the externally mediated fashion of linear sequence. Instead the relation is one of *internal* mediation, of the *mutual permeation* of opposites. Perspectives are grasped as flowing through each other in a manner that blends space and time so completely that they are no longer recognizable in their familiar, categorically dichotomized forms. You can see this most readily in viewing the fleshed-out cube. When you pick up on the odd sense of self-penetration of this seemingly impossible figure, you experience its two modalities neither simply at once, nor one simply followed by the other, as in the ordinary, temporally broken manner of perception; rather, you apprehend the *dynamic merging and separating* of perspectives. This, I submit, is what Schwartz-Salant was adumbrating when he spoke of the subtle body as entailing a space "in which 'outer' and 'inner' are alternatingly both distinct and the same."

Nevertheless, the Necker cube does have its limitations as a structure that could fully enflesh the subtle body. Schwartz-Salant portrays the subtle body as being no mere object in Cartesian space but as possessing a peculiar "subjective-objective quality."[4] While this merging of subject and object is indeed effectively *symbolized* by the merging of inside and outside suggested in the integrative way of viewing the cube, the concrete fact is that the cube itself appears as but an object cast before our detached gaze. The subjectivity of he or she who views the cube is not tangibly implicated in the subject-object merger that is depicted.

Underlying this disjunction is the factor of *dimension*. The basic cube (fig. 3.1b) is a one-dimensional line drawing embedded in two-dimensional space (the flat surface of the page). Therefore, though the perception of the cube simulates our experience with solid entities in three-dimensional space, the cube as such is certainly no solid. In the three-dimensional context, when the cube is taken literally rather than figuratively, it is thus just an object I reflect upon, one that appears before me, a being that is circumscribed, closed into itself, closed off from the inwardness of this three-dimensional subjectivity that does the reflecting. So the transpermeation of subject and object characteristic of the subtle body cannot fully be delivered by the dimensionally inadequate Necker cube. Is it possible that, if a higher-dimensional version of the cube were available, we would be able to go further?

I suggest the answer is yes. I propose that we could go beyond a merely symbolic representation of the subtle body if a solid, full-fleshed, three-dimensional counterpart of the cube were accessible to us. This paradoxical body could be experienced in such a way that while standing before me, it would also stand *within* me. It would present itself to me from the inner core of itself and I would recognize that core as my own. That is to say, the three-dimensional object and the dimension constituting my subjectivity would be utterly open to one another, would permeate each other in an unobstructed, boundless exchange. Thus, whereas we do not "lose our objectivity" in perspectively integrating the one-dimensional cube, whereas the integrated viewing of this object does not encompass our own subjective viewing process, the perspectival integration of a three-dimensional body would be different. Now integration of what is "out there" would be achieved through a process that would carry us back to what lies "in here," to our ownmost lived subjectivity, that is—to the sub-objectivity of the uroboros.

TOPOLOGY

We know that the subtle body is intimately related to the Hermetic vessel of alchemy. In my 1995 essay cited in the previous chapter, I attempted to express old alchemy's mystery vessel in modern terms by turning to the field of qualitative mathematics known as *topology*. Conventionally defined, this is the study of geometric objects that stay the same when they are stretched or deformed. But while convention adheres to the objectivist paradigm, we are going to employ topology in the *sub*-objective fashion of alchemy. It is topology that will permit us to convey the Necker cube experience of perspectival integration in a more palpable, higher-dimensional form, thereby preparing the approach to a fuller realization of the subtle body. Let us begin our topological odyssey with a comparison.

A cylindrical ring (fig. 3.3a) is constructed by cutting out a narrow strip of paper and joining the ends. The surface of Moebius (fig. 3.3b) is produced simply by giving one end of such a strip a half twist (through an angle of 180°) before linking it with the other.

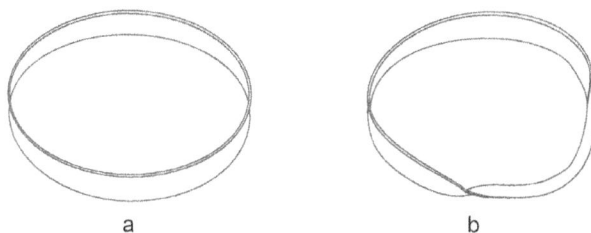

a   b

Figure 3.3 Cylindrical ring (a) and Moebius strip (b).

The cylindrical ring possesses the conventionally expected property of two-sidedness: at any point along its surface, two distinct sides can be identified. Now, in the Moebius case, it is true that if you place your index finger anywhere on the surface, you will be able to put your thumb on a corresponding point on the opposite side. The Moebius strip does have two sides, like the cylinder. But this only holds for the local cross-section of the strip defined by thumb and forefinger. Taking the full length of the strip into account, we discover that points on opposite sides are intimately connected—they can be thought of as twisting or dissolving into each other, as being bound up internally. Accordingly, mathematicians define such pairs of points as *single* points, and the two sides of the Moebius strip as but *one* side. If the Moebius property of one-sidedness is difficult to imagine in the abstract, it is very easy to demonstrate. Starting on one side of the strip, draw a continuous line along its whole length. Upon returning to your point of departure, you will discover that your ink mark has covered *both* sides of the surface—something that would not happen with a line drawn on the two-sided ring.

It is important to recognize that the surface of Moebius is not one-sided in the homogeneous sense of a single side of the cylindrical ring. It is one-sided in the paradoxical sense, one-sided and also two-sided, for the local distinction between sides is not simply negated with expansion to the Moebius as a whole. In coming to interpenetrate each other, the sides do not merely lose their distinct identities. Moebius oneness is essentially similar to the oneness of the perspectivally fused Necker cube. There is inside and there is outside. The two are different. Yet they also are one and the same.

The relationship between the Moebius surface and the Necker cube can be likened to that between a sculpture and a painting. The two art forms are both external representations of inner dimensions of experience (thoughts, intuitions, feelings). But the sculpture, by making significant use of three dimensions instead of being limited to two, can express the subject matter more concretely, flesh it out through the tactile sense as well as the visual. In like manner, since the Moebius strip is a two-dimensional surface embedded in three-dimensional space, it can embody the paradoxical union of opposites more concretely than can the lines of the schematic cube, limited as they are to a two-dimensional medium of expression.

Nevertheless, while the Moebius model manifests one-sidedness more tangibly than the cube, it *is* a model, an outward symbolization of the union of inside and out, rather than a full-fledged embodiment directly incorporating the inner depths of the subtle body. What would be needed for the latter? Not a two-dimensional body enclosed as mere object in three-dimensional space, but a body of paradox that is itself three-dimensional.

Let us return for a moment to the one-dimensional Necker cube (fig. 3.1b). Viewing its constituent perspectives as they are displayed sep-arately in Cartesian fashion (fig. 3.1a), you already may have noticed that each of these components of the reversible cube is also reversible: you can view either perspective in either a concave or convex fashion. Moreover, from the Necker cube that integrates opposing Cartesian per-spectives, a new order of opposition arises.

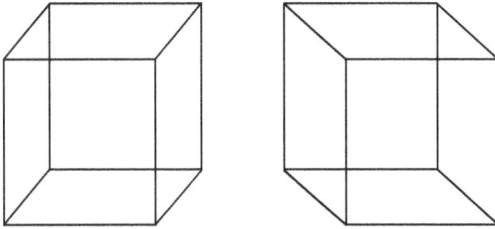

Figure 3.4 Necker cube opposites.

Figure 3.4 shows that the Necker cube itself possesses a mirror oppo-site with which it presumably could be integrated to create a third order of self-reversal. There is an unmistakable dialectical pattern at play here wherein the same integrative action that surpasses simple opposition at one level brings new opposition at another. We may study this pat-tern more effectively in the denser, more concrete medium of the two-dimensional surface.

Above I demonstrated the construction of the two-sided cylindrical ring (fig. 3.3a) and one-sided Moebius strip (fig. 3.3b). Mathematicians have investigated the transformations that result from *bisecting* topo-logical surfaces. Suppose we were to cut the cylindrical ring down the middle, proceeding along its full length. Upon completing the cut, the ring would simply decompose into a pair of identical narrower rings each possessing the same topological structure as the original. A more inter-esting result is obtained in bisecting the one-sided Moebius strip. When the operation is completed, instead of falling into two separate pieces as one might expect, the bisected surface retains its integrity but has now become a *two*-sided structure. We may compare a single side of this two-sided surface with that of the two-sided cylindrical ring.

We see from figure 3.5 that, whereas revolution about the cylindri-cal ring (right) describes but one closed loop, traversal of the bisected Moebius strip (left) gives us a doubly-looped, figure-8 pattern known in mathematics as a *lemniscate*. With the lemniscate, the familiar sign for infinity (∞) is rendered topologically.

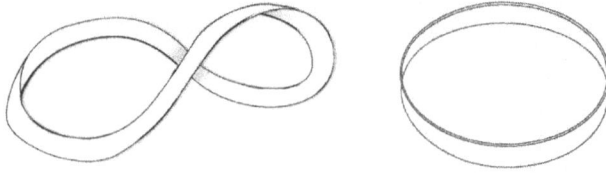

Figure 3.5 Lemniscate (left) and cylindrical ring (right).

Now, we have found that the integrative quality of the Moebius surface lies in its paradoxical one-sidedness. The two sides of the Moebius flow unbrokenly into each other to form a single side, without either side actually losing its distinctness. When bisection of the Moebius strip transforms it into a *two*-sided structure, this integrity is lost. Yet we can now see that each of those sides, being lemniscatory in character, constitutes its own order of paradoxical "unity-in-diversity." That is, in a single side, we have a double cycle, these cycles being connected by a continuous movement through the central node of the infinity figure (∞) that reverses rotational direction from clockwise to counterclockwise. It is true that, in the lemniscate, we no longer have a complete overlapping of opposing elements, as we do in the Moebius. The lemniscate thus could be said to have less internal coherence than the Moebius. Be that as it may, a similar pattern of "transpolar flow" is evident in both structures.

The relationship between the lemniscate and Moebius surface is analogous to that between the first- and second-order cubes of figure 3.1. Just as each first-order cube of figure 3.1a constitutes an internal self-reversal in its own right, each of the sides of the bisected Moebius strip is composed of the integrated cycles of the lemniscate entailing a reversal of rotational direction. And just as fusion of the externally related first-order cubes gives the second order of cubic self-reversal (fig. 3.1b), fusion of the opposing lemniscatory sides (by gluing together the edges of the bisected Moebius along the seam to undo the bisection) restores the one-sided Moebius (fig. 3.3b) in its original integrity. Furthermore, similar to the way a still higher order of integral self-reversal is intimated in the dialectic of the cubes (fig. 3.4), a new order of integration is indicated for the paradoxical surfaces.

Like the Necker cube opposites, our corresponding topological surfaces are *enantiomorphic* in nature: they possess the asymmetric quality of mirror opposition. Thus, the single lemniscate has its mirror counterpart on the other side of the bisected Moebius strip, and the Moebius as a whole has its own enantiomorph. Whereas there exists but one form of the symmetric cylindrical ring (fig. 3.3a), the Moebius surface can be produced in either a left- or right-handed version, depending on the direction in which it is twisted. If left- and right-oriented Moebius strips were stretched out and glued together edge to edge, a topological structure called a *Klein bottle* (fig. 3.6) would result (named after the German mathematician Felix Klein).

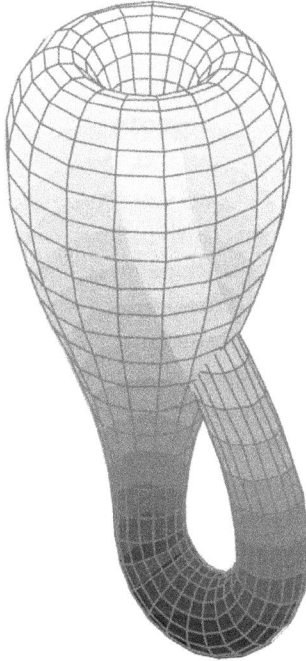

Figure 3.6 Klein bottle. (Courtesy of Tttrung, Wikipedia.org)

The Klein bottle has the same property of asymmetric one-sidedness as the two-dimensional Moebius surface, but implicates an added dimension.[5] Mathematicians tell us that we cannot really produce a proper physical model of this curious bottle. That is, left- and right-facing Moebius bands cannot be joined together in three-dimensional space without *tearing* the surfaces. I am going to suggest that this inability to objectify the Klein bottle in three exterior (Cartesian) dimensions derives from the fact that the bottle actually calls into play the inside-out *uroboric* dimension!

There is a different but mathematically equivalent way to describe the making of a Klein bottle that, for our purposes, will be very instructive. Once again a comparison is called for.

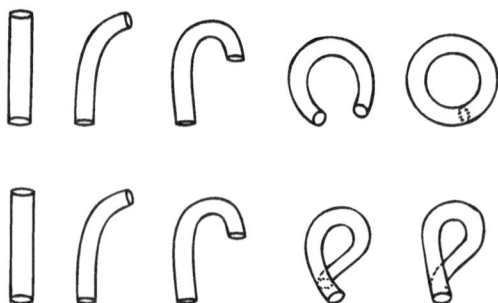

Figure 3.7 Construction of torus (upper row) and Klein bottle (lower row).

Both rows of figure 3.7 depict the progressive closing of a tubular surface that initially is open. In the upper row, the end circles of the tube are joined in the conventional way, brought together through the three-dimensional space outside the body of the tube to produce a doughnut-shaped form technically known as a *torus* (a higher-order analogue of the cylindrical ring). By contrast, the end circles in the lower row are superimposed from *inside* the body of the tube, an operation requiring the tube to pass *through* itself. This results in the formation of the Klein bottle. Indeed, if the structure so produced were cut in half, the halves would be Moebius bands of opposite handedness. But in three-dimensional space, no structure can penetrate itself without cutting a hole in its surface, an act that would render the model topologically imperfect. So, from a second standpoint, we see that the construction of a Klein bottle cannot effectively be carried out when one is limited to the three Cartesian dimensions that frame our experience of external (objective) reality.

Mathematicians are aware that a form that penetrates itself in a given number of dimensions can be produced without cutting a hole if an *added* dimension is available. The point is nicely illustrated by mathematician Rudolf Rucker.[6] He asks us to imagine a species of "flatlanders" attempting to assemble a Moebius strip.[7] Rucker shows that, since the "physical" (externally experienced) reality of these creatures would be limited to *two* dimensions, when they would try to make an actual model of the Moebius, they would be forced to cut a hole in it (fig. 3.8). Of course, no such problem of the Moebius intersecting itself arises for us human beings, who have full access to three external dimensions. What is problematic for us is the making of the Klein bottle, requiring as it would a fourth dimension. Try as we might, we find no such dimension "out there" in which to execute this operation. The critical distinction between the standard mathematical interpretation of the Klein bottle and a modern alchemical one hinges on their different ways of dealing with the fourth dimension.

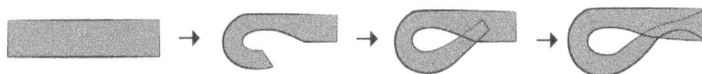

Figure 3.8 Construction of a Moebius strip in flatland: since the operation is limited to two-dimensional space, the form would have to be cut open at the place where it intersects itself (see third stage). (Adapted from Rudolf Rucker, *Geometry, Relativity, and the Fourth Dimension* [New York: Dover, 1977], 54, fig. 76)

Beginning on a more general note, we can say that the crux of the difference lies in the fact that while conventional mathematics tacitly upholds the split between psyche and physis, alchemy seeks to surpass it. In the conventional approach, the concrete subjectivity of the psyche is denied and attention is limited to the mathematical object. If an object requires three or fewer dimensions for its proper construction (as in the case of the cylindrical ring, Moebius surface, and torus), a "real" (i.e., tangibly perceptible) model of it may be successfully fashioned; if the object cannot be assembled in three-dimensional space, objective reality may be extrapolated, extended by an act of abstract imagination in which one or more additional dimensions are summoned. In this view, the fourth dimension required to complete the Klein bottle remains essentially an external dimension, albeit an imaginary one, and the Klein bottle is taken as an "imaginary object." Therefore, whether the mathematical object can be concretized or must be approached through abstraction, the conventional mathematician's attention is always directed outward toward an object; it may take passing note of its own subjective operations but never will make them its primary focus. Or we may say that mathematical awareness traditionally is geared to move *forward*, to project objects before itself as finished products without paying heed to the underlying process by which those projections occur.

The old formula of object-in-space-before-subject clearly is at play in the customary mathematical treatment of the Klein bottle. It is this that would be challenged in an *alchemical* mathematics. Our look at Western history has told us that the post-Renaissance preoccupation with an objectified physis observed from the detached viewpoint of an abstracted psyche may be ready to yield, in preparation for the concrete reunion of physis and psyche found in alchemy. In what specific form appropriate to the modern alchemical context will the integration of psyche and physis become manifested? My proposition is that it will be expressed as a *dimensional* integration. That is, physis, the dimension convention-ally associated with extensive, objectified three-dimensional space, will become integrated with a "fourth" dimension. The latter will not be a dimension that is simply extended before us in the manner of an exter-nalized space, as in the standard mathematical approach. Rather, physis will merge with the *in*tensive dimension comprising our thoughts, feel-ings, sensations, and intuitions—the whole of our subjectivity. Gaining access to these dimensional intensities will require the switching of gears that we explored in the previous chapter. To incorporate the dimension of the psyche, the mathematician as alchemist will need to move backward against the prevailing grain of Projective activity in acts of Propriocep-tion that will withdraw those Projections. By thus moving backward as we continue moving forward (Projective activity will not simply cease), the new alchemy will consummate the marriage of psyche and physis in completion of the subtle body.

    Again, it is a question of properly fashioning the alchemical vessel. The bottle must be hermetically sealed, closed so tightly that it also is open; the outer surface of the vessel must be finished in such a way that what lies within it is set free—not merely by being separated from the containing surface but by being integrated with it through a process of *self*-containment that brings to fruition the bottle's uroboric inside-out nature ("the outside to the inside, the inside to the outside").[8] Jung's description of alchemical distillation and the vessel in which it occurs (already noted in chapter 2) attests to their "backward orientation": the distillation "was in some way turned back upon itself" and took "place in the vessel called the Pelican where the distillate runs back into the belly of the retort" (see fig. 2.3).[9] Of course, the merged "sides" of the com-pleted vessel will at bottom not be just the sides of some object appearing out in space, some entity observable to a subject who remains detached from it. Objectivity and subjectivity themselves will be the "sides" that will permeate one another.

    What the modern alchemist can grasp that the ancient alchemist could not is that the merger of objective and subjective sides entails precise-ly the interpenetration of extensive, external three-dimensionality with our fourth, intensive, interior dimension. Just this *coincidentia opposi-torum* is embodied in the one-sided Klein bottle, understood in the light of alchemy. We have seen that there is a sense in which the Klein bottle does have the quality of an object in space insofar as a palpable model of

it can be approximated in three exterior dimensions; but we know as well
that the bottle is not simply objectifiable, for it is not an entity like the
torus whose construction in three-dimensional space is unambiguously
completable. This is because the Klein bottle penetrates itself, and, when
it is confined to three exterior dimensions, the penetration can only be
achieved by the topologically impermissible operation of cutting it open.
Yet we have also discovered that the opening which appears in the objective three-dimensional model of the bottle is filled by introducing a
"fourth" dimension. Interpreting that dimension alchemically, viewing
it as the inner dimension of psyche, the Klein bottle becomes identified as the present-day counterpart of the Hermetic vessel (as even its
outward appearance seems to suggest; see fig. 3.9). The new incarnation
of the bottle, being made of "perfected glass"—constructed in terms of
the conceptually mature, highly differentiated idea of topological dimensionality—can contain the "Mercurial wine" in a way the old bottle
could not. When the sides of the Klein bottle fuse, they do so without
con*fusion, without losing their distinctiveness (this was also noted for
the sides of the Moebius surface, the bottle's lower-dimensional equivalent). By means of the Klein bottle, outside and inside, physis and psyche,
are sealed off from one another in such a way that, paradoxically, they
totally mesh.

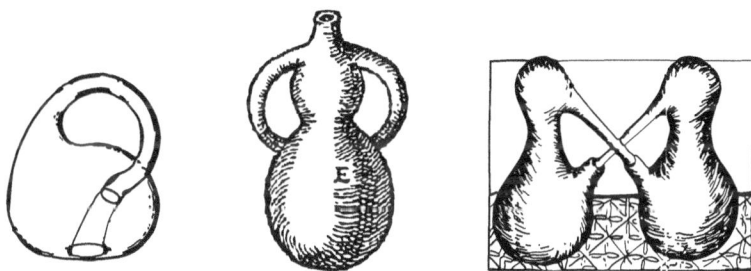

Figure 3.9 The Klein bottle (left) and images of the Hermetic vessel (center and right).
(Image of Klein bottle is from Martin Gardner, *The Ambidextrous Universe*
[New York: Charles Scribner's Sons, 1979], 151)

Let me emphasize that the Klein bottle, as Hermetic vessel, seals outside and inside from one another *hermetically*; that is, opposing sides are
actually differentiated more completely than in conventional structures.
The implication here is that the post-Renaissance division of object and
subject is by no means complete. In viewing an object in the conventional way, we naively assume that it is a self-standing, simply autonomous
entity by neglecting our own subjective role in the process. Both in con-

temporary science (especially theoretical physics) and in contemporary philosophy (especially phenomenology), recognition has come that observer and observed, subject and object, are so intimately interconnected that, in effect, they are inseparable (see my *Dimensions of Apeiron* and *The Self-Evolving Cosmos*). This makes it naive indeed to continue thinking of the observed object as being merely "out there," standing on its own in simple autonomy from the observing subject "in here." By continuing to assume a separation that does not exist and can never be realized, the process of *authentic* separation is obstructed.

This idea is nicely brought out in Jung's concept of individuation. From the Jungian standpoint, the goal of development is to make the unconscious conscious, to overcome defensive denial and confusion about oneself and gain full-fledged self-knowledge. What the individual is not consciously aware of in herself makes its presence felt *un*consciously, in a deficient, undifferentiated form. Now, the *most basic* impediment to individuation lies in our most basic misconception: we mistakenly view ourselves as isolated, simply self-subsistent individuals. For Jung, and indeed, for alchemy in general, one completes one's development as a distinct individual only in fully recognizing the truth of one's intimate entwinement with others, with nature, and with the cosmos as a whole. Therefore, being fully individuated means being fully integrated. Whereas the conventional dualism of post-Renaissance experience achieves neither of these aims, the new alchemy would realize both.

And precisely this thoroughgoing differentiation and integration of subject and object is found in the "new alchemical vessel," embodied in the Klein bottle's transpermeation of intensive and extensive dimensions. Without regressively dissolving, each side of the bottle expresses itself *as* the other side, and, in so doing, expresses its own individuality more truly and fully than in conventional structures such as the torus, which lend themselves to the dualistic "separation" of object and subject that obscures their underlying relatedness. The paradox then is that, while the sides of the Klein bottle are wholly identified with one another, they are also wholly themselves, thus wholly differentiated from each other. This is what it means for alchemy's vessel to be *bene clausum*, "hermetically sealed."

What was implied in the previous chapter is that the vessel must be sealed *more than once*. Completing the vessel in glass symbolizes full-fledged intellectual maturity, the stage of development that Renaissance alchemists called the *unio mentalis*. To repeat what Jung said on this,

"The [glass] bottle is an artificial human product and thus signifies the intellectual purposefulness and artificiality of the [alchemical] procedure."[10] But Jung also notes that this "mental union" is "only the first stage of conjunction or individuation" and that further distillations are required, processes entailing a reunion of mind with body, and with the whole of nature.[11] Indeed, we found in chapter 2 that the uroboric vessel must ultimately be sealed in *stone*, rendered as the Philosopher's Stone.

In the chapters that follow, we will explore in detail the several stages of alchemical transformation and their associated forms of topological containment. As this odyssey unfolds, we are going to discover that (1) with each incarnation of the vessel, a new psychophysical dimension is brought into play; that (2) access to each dimension requires a movement Proprioceptively backward and downward into a denser realm of the underworld; and that (3) each hermetic closure involves an alchemical ordeal culminating in death and resurrection—the demise of a finite egoic body and its rebirth as infinite subtle body.

CHAPTER FOUR

The *Unio Mentalis*

Opening Stage of Alchemical Conjunction

SOLVE ET COAGULA: A PRELIMINARY OVERVIEW OF THE STAGES

As it happens, we will require more than the Klein bottle to do justice to the topological intricacies of alchemical process. To begin to see why this is so, let us reconsider the motto of alchemy introduced in the preface: *solve et coagula*, "dissolve and coagulate."

The verb *dissolve* (from the Latin *dissolvere*) indicates separation: to dissolve is "to disunite; to break up . . . to cause to separate into parts" (*Webster's [Unabridged] Dictionary*). The alchemical substance to be transformed was initially broken into parts, sublimated so as to refine it. Then the process was reversed and the material was made to congeal, its constituent elements being drawn together and solidified. This was the phase of *coagulation* (from the Latin *coagulatus*, past particle of *coagulare*, "to curdle"). What had been divided for the purpose of refinement was now reconstituted.

The dictum of alchemy makes it clear that, before there can be a conjunction, there must be a *dis*junction. Bearing in mind that alchemy was not just concerned with transforming chemical substances in a laboratory beaker but with the transformation of psyche and physis in general, we may broadly say that the *solutio* is characterized by the transition from a state of concretely embodied polymorphous relatedness to one of monomorphous abstraction: a single order of being gains ascendancy, detaching itself from the opposing orders with which it had been intimately entangled, whereupon the latter are objectified or repressed. We have seen this process at work in the forward thrust of Greek patriarchy out of the uroboric womb of mythic culture. The separation was achieved in such a way that the polymorphous goddesses and gods of the older world were displaced by the solitary God of monotheism. The *solutio* is also evidenced in the post-Renaissance Projection of object-in-space-before-subject, wherein the subject is the transcendent ego that has risen above its worldly participation with other beings and now takes these others as objects appearing before it in empty space. We recognize this ego as the Cartesian *cogito* or thinking subject, the mental being that has abstracted itself from its body and has objectified that body. Phylo-genetically, the *solutio* is expressed in humanity's loss of contact with the rest of nature. The close connection with the animal, vegetable, and mineral spheres palpably felt in mythic and indigenous cultures gives way to an anthropocentric domination of the planet. And deeply related to the *solutio* is the change in attitude toward death discussed before: the primal transpermeation of life and death dissolves into an abstraction of life in denial of death.

What happens in making the transition to the *coagulatio*? We turn back to the intricately interwoven polymorphous complexity that presents its challenges for topological containment—challenges that can only be addressed by going beyond the Kleinian container. But we are not yet ready to confront these challenges. We first need to clarify further the several specific stages of alchemical conjunction, beginning with a reconsideration of the opening stage, the *unio mentalis*.

Citing sixteenth-century alchemist Gerard Dorn, Jung says: "The *unio mentalis*, the interior oneness which today we call individuation, [Dorn] conceived as a psychic equilibration of opposites 'in the overcoming of the body,' a state of equanimity transcending the body's affectivity and instinctuality."[1] Here,

> the mind (*mens*) must be separated from the body. . . . By this separation (*distractio*) Dorn obviously meant a discrimination and dissolution of the "composite," the composite state being one in which the affectivity of the body has a disturbing influence on the rationality of the mind. The aim of this separation was to free the mind from the influence of the "bodily appetites and the heart's affections," and to establish a spiritual position which is supraordinate to the turbulent sphere of the body.[2]

Therefore, in achieving the intellectual maturity that brings mental development to fruition, one must surmount "everything bodily, sensuous, and emotional . . . the soul's appetites and desires."[3]

Yet Dorn well recognized that the *unio mentalis* is only the *opening* stage of conjunction. Stage two of the process occurs when the mind, having effected its separation from the body (*solve*), is then reunited with it (*coagula*). Does this mean that the *unio mentalis* as such is a *coniunctio* without being a *coagulatio*? Is it a purely disembodied mental union exclusively associated with the *solutio*? I suggest that the meaning of the *unio mentalis* is subtler than that. We have found that the *solutio* entails an ascendancy of mind in denial of the body. But the blindness of mind to its origin in the body bespeaks a mental limitation. In reaching the genuine fulfillment of the *unio mentalis*, the mind must consciously recognize its bodily origin and the necessity of returning to that body in order to carry its individuation even further. What we will discover before this chapter is done is that the *unio mentalis* culminates with its own subtle order of embodiment.

Note that, when Dorn speaks of overcoming the body via the *unio mentalis*, two spheres of embodiment are specified: "the bodily appetites and the heart's affections." Jung apparently associated these spheres with "sensuous" and "emotional" functioning, respectively.[4] Presumably, in the second stage of conjunction, when the mind reunites with the body, the bodily centers in question, having been subdued in the *unio mentalis*, would now be reanimated. I propose that what Jung (following Dorn) identified as the second stage be seen as actually entailing *two* stages, the first of these involving emotion and the second sensuality. This accords with a fundamental distinction made by Jung himself, and we are going to see that such a formulation permits the alchemical stages to be coordinated with the topological ones.

In *Psychological Types*, Jung maintained that the functioning of the human psyche entails four basic forms of activity.[5] These functions cannot be reduced to each other and are invariant relative to the specific content of experience, which changes from moment to moment. The four functions of which Jung spoke are *thinking, feeling, sensing,* and *intuition*. Jung viewed thinking and feeling as diametrically opposed to each other and as "rational." That is, both are *representational* functions; they involve a reflection on, and an (e)valuation of, externally experienced reality from the inner perspective of the subject. In the case of thinking, the subjective base is abstract and disembodied (e.g., discursive reasoning, logical deduction, mathematical calculation), while, in the case of feeling, it is more concrete and embodied (I value what lies outside me in terms of the likes and dislikes of my body).

Beyond these "rational" operations lie the "irrational" functions of sensing and intuition, also seen to make up a pair of opposites. Irrational activity can be characterized as more "presentational" than representational; that is, it entails a more immediate (less reflective) reaching out to, grasping, and taking in of that which is other in the field of experience. In sensory perception, we experience the world discretely, dividing the objects we encounter into units that are bounded from each other. On the other hand, intuiting the world means apprehending it as an undivided whole, as in cases of "hunches" or "visions," where we are seized by a nebulous impression about the general course of events but are unable to account for the source of the presentiment. While the irrational functions are attenuated in adulthood, being constricted by the ascendant influence of rational thinking, according to Jung these functions originate in an "infantile and primitive psychology" that serves as the "matrix out of which thinking and feeling develop as rational functions."[6] Evidently, intuition would be the *most* primal form of experience, for while sensation involves "perception via conscious sensory functions," intuition entails "perception by way of the unconscious," through "dreams and fantasies," through a mode of operating in which one surrenders oneself "wholly to the lure of possibilities."[7]

Implicit in Jung's analysis of psychic functioning is a developmental sequence that lends itself to alignment with the alchemical stages. The emergence of the thinking function from the "infantile and primitive" matrix of the more concrete functions clearly corresponds to the *solutio* that culminates in the *unio mentalis*. In the stages of conjunction that follow, the feeling function—the "heart's affections," as Dorn put it— would be the first to be "let out of the bottle." The third stage of conjunction would bring liberation of the more concrete sensing function which, in its primal form, involves the "bodily appetites," including instinct and sensuality. Finally, there would come the stage of the *coniunctio* that would take us to the deepest and darkest stratum of the psyche. This would involve the functioning of primordial intuition governed largely by the unconscious, that realm of pure "possibilities." The *unus mundus* is the realm I am speaking of—"the potential world of the first day of creation, when nothing was yet 'in actu.'"[8] Von Franz associated this with the "realm of the dead."[9] Citing Dorn, Jung is explicit on the developmental significance of engaging with this dark sphere: "Dorn sees the . . . highest degree of conjunction in a union or relationship of the adept . . . with the *unus mundus*."[10] We discovered in chapter 2 that said *coniunctio* would seal the Hermetic vessel in "stone." That is to say, it would be a realization of the Philosopher's Stone that would mark the completion of the subtle body and would thus cement the resurrection of the ego—

not in a finite particular form but as a "universal ego," as the uroboric Self.[11]

What of the *topological* containment of these *coniunctios*? Would the Klein bottle suffice to encompass them all? Above I intimated that it would not. That is because, in descending from the *unio mentalis* into the more densely embodied conjunctions, whole new dimensions open up requiring distinct forms of topological support. But before we can explore in depth the denser conjunctions and their topo-dimensional correlates, we need to go further by way of bringing to completion the opening conjunction. What is called for is a *self-signification of the text.* The remainder of the chapter will be devoted to this.

THE *UNIO MENTALIS* AS A SELF-SIGNIFICATION OF THE ALCHEMICAL TEXT

Merely writing about the *coagulatio* as I have will not bring it to full-fledged actuality. This does not mean, of course, that the text should simply be abandoned, since the alchemical opus we are embarked on is not just an enterprise involving physical substances in a laboratory beaker, but a *psycho*physical endeavor, one featuring a significant conceptual aspect that can be brought out through the written text. So, if writing about the *coagulatio* as I thus far have done will not by itself bring alchemical conjunction to fruition, might there be another mode of writing more likely to facilitate the task?

The fact is that my manner of writing this text has been largely at odds with the *coagulatio*. I am not referring to the explicit content of what I have expressed but to the form my expression has taken. By and large I have tacitly adhered to the post-Renaissance default setting that governs writing wherein the text moves out of itself in such a way that, in delivering a semantic content for reflection, its pre-reflective delivery process is obscured. Insofar as attention has not been called to the concrete acts of signification in which I have engaged—to the writing process from which my meanings are produced—the meanings thus abstracted appear as free-floating semantic objects. I may write, for example, of a *unio mentalis* that goes beyond the Cartesian splitting of subject and object, but this signified content is implicitly undermined in the way it is presented: having been detached from the act of signification from which it derives, it is offered in such a way that it is itself but an object, with the subject (the author and his writing process) relegated to the background. (While there are aspects of this book that do shed some light on my process, their mode of expression lacks sufficient concreteness, as will become more evident in further clarifying the meaning of self-signification.) Therefore, if this text is to enact a bone fide *coagulatio*, it cannot limit itself to presenting semantic products as does the post-Renaissance text. In signifying alchemical conjunction the text must act recursively to signify *itself*, its own means of action. Through such self-signification, we would not just be writing about an alchemical conjunction happening somewhere "out there" beyond the text; the conjunction would be happening *here and now, in this very writing*. What we need to do then so as to enact the *coagulatio* with sufficient concreteness is switch from merely writing *about* conjunction to a conjunctive self-writing. But how can this be carried out? I will address the question via semiotics, the study of signs, signifiers, and the process of signification.

THE KLEINIAN SIGNIFIER

The classical text of the post-Renaissance era operates squarely within the reflective mode and raises no questions about doing so. Here the word or sign, whose signifier serves as surrogate for the subject, refers solely to what is other, making this signified object of reflection explicit, a well-bounded content closed into its context (an object in semiotic space). The signifier/subject per se remains entirely *im*plicit; it does not meaningfully refer to itself. Is classical reflection effectively challenged in modernist or postmodern deconstructionist writing? I suggest that it is not.

It is true that the modernist sign is self-referential. We can say that, in modernism, attention is withdrawn from the end-products of reflection and meaning is relocated in an abstraction of the process itself. We see this in intensely self-involved psychological works such as James Joyce's *Ulysses*. In fact, we may say that modernism as such is "psychological": it seeks to apply logos to the psyche, that is, to gain explicit knowledge of subjectivity. Here the sign is turned back upon itself so as to bring to light what formerly had been strictly implicit. In the language of Freudian psychoanalysis—that exemplar of modernism—the goal is to "make the unconscious conscious." Stated most essentially, modernism wants to surpass classical signification by turning the classical subject into an object.

In carrying out its program, does modernism achieve its basic goal? What actually happens is that the classical approach is upheld at a higher level of abstraction (see the discussion of modernism in chapter 2). For, in making the old subject explicit, in rendering it a well-delineated content enclosed in its context, this object-née-subject implicitly must be given to, must appear before, a new, higher-order subject.[12] The self-reflection of modernism is indeed akin to gazing at the reflection of one's eyes in a mirror: what *had* been the gaze of the subject now itself appears as an object gazed upon by a subject that is one step removed from the original. The important thing to recognize is that this transformation of terms leaves completely intact the classical *relationship* of object-in-space-before-subject. The move to modernism thus poses no fundamental challenge to the mode of signification that preceded it. Modernism's self-referential sign, by turning the self into an other, in fact maintains in abstraction the classical *split* between self and other, between the signifier and its signified object.

In postmodernism, the ultimate consequence of modernism is recognized and played out. The crux of postmodernism, I suggest, is the realization that modernism's objectification of the subject is but the first term of an infinite regress. It might seem that, in modernism, only the *classical* subject loses its privileged position as the unquestionable base of knowledge, as the transcendent, never-to-be-viewed perspective point from which all else is viewed. However, once this subject is viewed, made explicit, objectified, cast before the perspective point of a newly implied, higher-order subject, *no* subject can securely hold its position. Having established that the classical subject can be turned into an object, the new, modernist subject should be susceptible to the same fate. The objectification of that subject would bring a still newer order of subjectivity with the same susceptibility, and so on, ad infinitum. And each time the subject is undermined by being made into an object, what had been object to that subject is also undermined. Ultimately then, we have neither subject nor object in any stable, abidingly meaningful form.

In the parlance of postmodern or post-structuralist literary theory, the fixed relationships between particular signifiers and their signified meanings give way to the restless, ever-shifting text, the ceaselessly self-alienating application of the sign to itself. This is evident in the writings of Jacques Lacan, for whom the sign dissolves in a never-ending series of displacements and slippages wherein "the subject disappears."[13] Similarly for Jacques Derrida: "sign will always lead to sign, one substituting the other ... as signifier and signified in turn."[14] In Derrida's own words, language must be understood as a field "of *freeplay*, that is to say, a field of infinite substitutions" in which identity fragments into sheer difference (*différance*).[15] I am proposing that the specific way this takes place is by a recursive process of self-referential mirroring in which, time and again, the signifier/subject is displaced by being made into the signified object of a newly implicit subject. Therefore, if we view the self-reflection of modernism as a mirroring that maintains in abstraction the classical relation of object-in-space-before-subject, postmodernism would constitute an infinite repetition of this mirroring, one that upholds classical identity in such a way that, in the end, it also negates it. The sterility of thus maintaining the classical posture is not lost on the postmodernists. In their abstract self-reflections, there is a distinct mood of disenchantment. Yet, because the postmodern writer can find meaning nowhere else, the rule of classical reflection lingers on, albeit in a thoroughly self-subverting manner.

A phenomenological alternative to modernism and postmodernism alike is intimated by philosopher-psychologist Eugene Gendlin. Gendlin offers us a text that can reflect upon its bodily, *pre*-reflective source. "Speaking is a special case of . . . bodily living," says Gendlin. "Our bodies perform the implicit functions essential to language. . . . Our bodies imply our . . . linguistic meanings."[16] Moreover, when we speak or write, the pre-reflective source of this activity is not simply left behind; it continues to operate in the very midst of our linguistic functioning. Thus, for example, "the most sophisticated details of a linguistic situation can make our *bodies* uncomfortable."[17] Could we not reflect explicitly upon the pre-reflective source of our reflection? Let us attempt such an act of self-reflection here, with the very words on this page. If Gendlin is correct, our reading of these words arises from our bodies, and since this bodily source goes on functioning even as our words now turn back upon it, it seems we should be able to realize that source in a bodily way so that our words no longer appear as mere abstractions. This is what Gendlin means when he proclaims that "words can say how they work": they work from the body, and, becoming cognizant of their own bodily underpinning, they can link back to it.[18] As in Gendlin's approach, the words of the postmodern text do reflect upon themselves, but only as disembodied signs ultimately devoid of meaning. Gendlin points us beyond postmodernism. In Gendlin's form of self-reflection, the text is not merely conscious of itself *as* an abstract text, but calls attention to the concrete process from which it originates. Only by gaining access to this pre-reflective "subtext" can we supersede the old post-Renaissance trichotomy of object-in-space-before-subject and signify the Self concretely. In Gendlin's terms, the pre-reflective is "pre-separated"; that is, it "comes before," is more primordial than the divisions arising in classical thinking and perpetuated in modernism and postmodernity.[19]

But a further step seems necessary if we are to close the gap between our reflection upon the pre-reflective and the pre-reflective itself. A *Kleinian* rendition of Gendlin's text is required, I suggest.

First let me emphasize that our post-postmodern text must be *paradoxical* in character. Again, what we are seeking to do is include in this alchemical reflection of ours the pre-reflective source of our activity, the uroboric "subtext" that normally is kept implicit. Our text—whose signifiers stand in for our*selves* (for me, who writes these words, and for you, who reads them)—is to draw back in upon itself, make reference to itself without alienating itself, as happens with modernist and postmodern texts; in so signifying itself, this text cannot merely turn itself into an other that is cast before a newly implied, more abstract self. Does this mean that the self that is signified must be the *same* self that is doing the signifying? Not exactly. If the self in question were *simply* the same, our reflection would collapse into mere self-identity. As long as we are engaged in reflection, are working with a text, are writing or speaking and not merely remaining silent, there can be no simple self-identity. Yet, even though the very act of reflecting upon the self turns it into what is other, it is possible for this other to flow right back into the source from which it arises, rather than appearing *merely* as an other cast before a new self. Thus, the self-reflection I am describing would give us neither self nor other, in the strictly oppositional sense of these terms. We would realize instead their paradoxical interpenetration. Just this *coniunctio* of self and other, of subject and object, is what we require to supersede the supremacy of reflective predication. Signifier and signified would be more than reciprocally interdependent in such a self-reflective text. They would be identified, utterly one. Yet they also would be two. By virtue of the latter aspect, reflection would continue; by virtue of the former, the "pre-separated," pre-reflective dimension of the Self would be brought into play.

However, for the gap between reflection and the pre-reflective to be closed in this way, the paradoxical return of signification to itself requires a signifier that possesses *sufficient dimensionality*. It is here that the Klein bottle plays its crucial role. Our work with this structure clearly is no exercise in "pure mathematics" in which a mathematical object is signified by a definition or algebraic formula. For us, the Klein bottle is not merely a signified object; it is a *signifier*, one that, indeed, paradoxically signifies itself. Or, in the language of semiotics pioneer C. S. Peirce, the Klein bottle is a "Sign of itself" (see communications theorist Paul Ryan's exploration of the relationship between Peirce's "Sign of itself" and the Klein bottle).[20] It is the *dimensionality* of this enigmatic Sign that permits us to close the gap between the reflective and pre-reflective.

I raised the issue of dimensionality in the previous chapter. There I pointed out that neither the Necker cube (fig. 3.1b) nor the Moebius strip (fig. 3.3b) were of high enough dimension to give full expression to the uroboric subtle body, that only the three-dimensional Klein bottle could serve in that capacity. What presently concerns us is the Klein bottle's role as signifier.

It seems clear that the Necker cube and Moebius signify uroboric paradox more concretely than words alone could do. These words you are reading are composed of arbitrarily devised, conventionally agreed upon tokens that refer to their content in a merely external manner, whereas the cube and the Moebius iconically embody that paradoxical content. Related to this is the dimensionality of the signifier. The one-dimensional typographic marks appearing on the two-dimensional surface of this page obviously fall short of tangibly delivering the three-dimensional reality of the uroboric paradox they signify. In contrast, while the cube is also one-dimensional (i.e., a line drawing), it is directly suggestive of the three-dimensional perceptual world, and the Moebius strip is a two-dimensional embodiment of paradox. These two structures can therefore express more concretely than mere words the conjunction of opposites that constitutes the Self's body.

Yet even though the Necker cube and the Moebius strip both signify the *coniunctio* more tangibly than does the written word, they do not go so far as to signify *themselves* in the process. Consequently, owing to their still insufficient dimensionality, they fail to enact a full-fledged fusion of inside and out, subject and object, self and other; instead, the conjunction they indicate is but a higher-dimensional *other*. What is needed to signify the full depths of the subtle body is not a one- or two-dimensional body enclosed as mere object in three-dimensional space, but a body of paradox that is itself three-dimensional, namely, the Klein bottle.

It is the *hole* in the Klein bottle that marks its self-signification. In the last chapter we concluded that the topological bottle is the modern-day incarnation of old alchemy's Hermetic vessel (see fig. 3.9). We also know that the design of the subtle vessel is essentially that of the serpent or dragon that consumes itself by swallowing its own tail (fig. 1.1). To take itself in, the serpent must intersect itself, an operation requiring a hole (corresponding to the opening that is its mouth). The Klein bottle's hole is of this sort. It is not merely a hole in a container, but the hole produced by the act of *self*-containment that integrates the container with its contents, in this way giving *(w)holeness*. This requires an extra dimension, of course (see previous chapter). Unlike the imaginary fourth dimension of conventional mathematics, alchemy's added dimension does not just eliminate the self-intersection of the Klein bottle and render it simply continuous. Instead, an aspect of discontinuity remains. There is in fact a *dialectic* of continuity and discontinuity, of whole and hole.

In semiotic terms, Kleinian self-containment entails a self-*signification*. This is the paradoxical process that would close the classical gap between signified and signifier while at once leaving it open. It is the action that would hermetically seal the breach between our reflection on the pre-reflective and the pre-reflective itself while at the same time maintaining reflective distance. No doubt there must be a hole in the Kleinian text, a break or discontinuity. Yet the hole created in the paradox of self-signification would enact the dialectic of discontinuity and continuity that would also give us *wholeness*.

By way of clarifying, let me take some time to expand on what I mean by "Kleinian text." On an obvious level, it is constituted by these one-dimensional typographic marks you are reading; by these conventional signifiers that point abstractly to higher-dimensional Kleinian self-signification. Of course, the conventional squiggles by themselves do not give Kleinian self-signification in *concrete* terms; they do not provide the necessary coagulation of the text enabling the reflective signifier to actually reenter its own pre-reflective ground. The Kleinian text is enhanced, made more concrete, by an iconic image of the Klein bottle (fig. 3.6) that raises the dimensionality of the signifier: now the Kleinian signifier is two-dimensional. Yet this dimensional enhancement still does not go far enough in distilling the Kleinian text. Two-dimensional images of the Klein bottle do not suffice in bringing about Kleinian *self*-signification because the Kleinian vessel is a structure that engages *three* dimensions and realizes concrete self-reference via the "fourth" dimension—the psychophysical dimension encompassing our lived subjectivity. What we require in this text then is the manifestation of a three-dimensional Kleinian signifier that opens the way to the alchemical blending of psyche and physis.

One method for achieving an in-text 3-D effect is the production of an *anaglyph*, a composite photograph made up of two superimposed images shot in contrasting colors. This approach could be taken with the Klein bottle, but the picture would have to be viewed with special glasses whose lenses are of corresponding colors. Beyond the fact that the anaglyph would require special viewing equipment, there is a deeper issue. Granting that such a Kleinian signifier would suggest three-dimensionality, how effectively would it open itself to the "fourth" dimension?

The anaglyphic Kleinian signifier comes ready-made, in the sense that no conscious perceptual adjustment is required of the viewer for an effective reading of the image. This way of presenting the Klein bottle is conducive to continuing to view it as something "out there," detached from the viewer's subjectivity. Bearing in mind the need to enlist the "fourth dimension" for the full-fledged coagulation of the Kleinian text, let me describe a different way of producing a 3-D Kleinian signifier, one that may be better suited to Kleinian *self*-signification.

Three-dimensional perception can be simulated by a stereoscopic technique that does not require special glasses or any other optical equipment. For a simple example of this that is relevant to our present concern, let's return to the Necker cube (fig. 3.1b). We can in fact construct the cube stereoscopically.

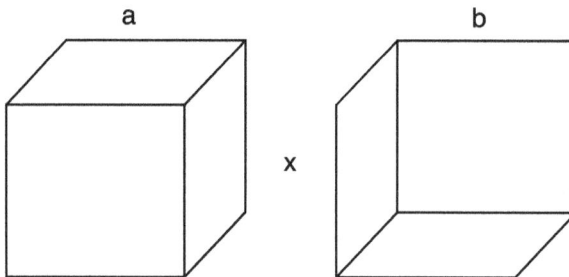

Figure 4.1 Stereoscopic construction of Necker cube.

Figure 4.1 parses the cube into its component perspectives. Instead of focusing on either perspective alone, direct your view to the "x" marked halfway between the perspectives and relax your eyes, allowing them to lose focus and cross. After a while, a third image should appear between the ones printed on the page. The new image fuses original perspectives, a fact you will be able to confirm by observing that the labels "a" and "b" have become superimposed. The image you have created stereoscopically is of course the Necker cube. In this stereogram, disparate images of the cube merge to yield an experience of three-dimensionality in which the cube seems to float off the page. Working with the cube in this manner should make it easier to work against the habitual tendency to fixate on just one of its perspectives, since you are now able to gain a proprioceptive sense of how both perspectives are integrally involved in the optical process by which the cube is produced. And this proprioception of the cube's production process surpasses the experience of the cube as but a ready-made object out in space, thereby heralding the *sub*-objective dimension.

While the one-dimensional (line-drawn) cube projects an image of three-dimensional reality, the higher-dimensional Klein bottle of course embodies that reality more fully. It is possible to construct a stereogram of the Klein bottle (fig. 4.2).

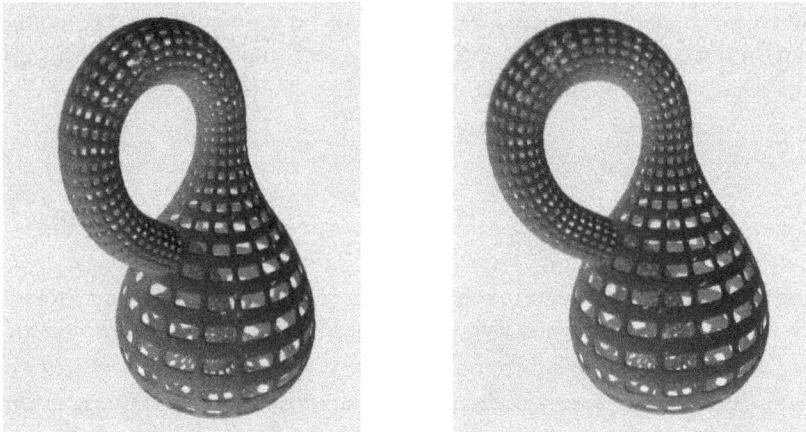

Figure 4.2 Stereogram of Klein bottle.

Once again, instead of focusing on either image of the figure alone, bring your attention to a point midway between the two images, relax your eyes, and allow them to cross. As with the cube, a third image will appear between the two that are printed, and this fusion of printed images renders the Klein bottle three-dimensional. Like the stereoscopic exercise with the Necker cube, the stereo-viewing of the Klein bottle entails a proprioception of the eyes in which your attention draws back in on your own optical activity as you bring into view a paradoxical structure that also reenters itself. In viewing this Kleinian signifier, what must be apprehended is its uroboric quality. As with the tail-swallowing serpent of old, the Klein bottle contains itself in such a way that it is at once both container and contained. If the Cartesian gap between subject and object is to be bridged, the Kleinian "object's" uncanny quality of self-containment is what must be digested as you continue to contain *yourself* by proprioceiving the action of your eyes. Of course, you can enact such proprioception in viewing *any* object, but an ordinary object will appear simply closed into itself, thus closed off from your subjectivity; the subject-object split will therefore be maintained. In proprioceiving your eyes as they gaze at the Klein bottle, it should be different. Because of its unique opening into the "fourth dimension"—the sub-objective, psychophysical dimension—the Klein bottle should be able to receive into its midst the proprioceptive subjectivity with which you view it. In this way, the optical proprioception you engage in while viewing the Kleinian stereogram should become intimately entangled with the Klein bottle's extra-dimensional return to itself. It turns out, however, that this stereoscopic practice alone does not take us far enough toward realizing our intention of surpassing post-Renaissance dualism and opening up the psychophysical dimension of the Self.

PROPRIOCEIVING THE "I"

Though I may engage in proprioceiving my eyes as I view the Kleinian stereogram and may register the Klein bottle's paradoxical self-containment, I continue to experience the bottle as an object "out there," an object in space cast before my still-detached subjectivity. Why is this so? It is because the subjectivity in question, that which needs to be Proprioceived, goes deeper than the action of my eyes. It is not just the eyes that must be attended to but the "I" that lies behind them. This is in keeping with the fact that the *unio mentalis* constituting alchemy's opening *coniunctio* entails more than a perceptual process. The process is primarily *cognitive*, for it is the *cogito* or thinking subject who must engage in self-signification. So the backward movement required has to go beyond the perceiving eyes to the conceptualizing "I"; this ego or core subjectivity cannot just proprioceptively *view* the Kleinian signifier but must Proprioceptively *read* it (the distinction between proprioception and Proprioception was first brought out in chapter 2). The upshot is this: as long as the abstract "I" currently presiding over this Kleinian text remains unacknowledged and unProprioceived, whatever "I" experience with "my" eyes will be experienced as but an object in space cast before "my" detached subjectivity, and the Cartesian gap will persist.

We may consider the Proprioception of the "I" while reading the

Kleinian signifier as involving the kind of nontraditional meditation touched on in chapter 2. Jung spoke of traditional Eastern and Western forms of meditation as being valuable "for increasing concentration and consolidating consciousness" but designed to "shield consciousness from the unconscious and to suppress it."[21] By contrast, the alchemical *meditatio* aims to augment knowledge of the unconscious and to effect a total integration of being in which the individuating ego realizes its grounding in the Self.[22] A critical part of this process is the withdrawal of one's projections: "If the individual is to take stock of himself," says Jung, "it is essential that his projections should be recognized."[23] This reversal of the forward thrust of one's projections is an aspect of what I have been calling *proprioception*. Just as one can obtain a propriocep-tive sense of the activity in one's eyes associated with the projection of images, one can gain proprioceptive awareness of the projections involved in the emotions, memories, and thoughts that make up one's personality.

Alchemy is of course not just concerned with retracting the particular ego's projections; it focuses on withdrawing the Self's own Projection of this ego per se. As a textual *meditatio*, this entails a Proprioception in which the self that is signified by the writer and reader of the text is in-deed the Self. Thus meditating Self-reflexively on the Kleinian signifier (fig. 4.2), the paradoxical content of what is signified becomes embodied in the Proprioceptive way the signifying is done, and the text quite nat-urally draws back in upon itself, moving backward through its hole to bring (w)holeness. In the serpentine *circulatio*, the Kleinian text twists radically around so that the text appearing before the Self at once lies within it ("the outside to the inside, the inside to the outside").[24] In this connection, recall Jung's characterization of alchemical imagination as "a hybrid phenomenon . . . half spiritual, half physical."[25] Applying said imagination to the dimensionally enhanced Kleinian body of this text, "the intermediate realm of subtle bodies comes to life again, and the physical and psychic are once more blended in an indissoluble unity."[26]

Still and all, if the *coagulatio* is to be fully realized, the Self cannot just withdraw the Projection of the ego in the abstract. For the conjunc-tion to take place right here in this text, the Projection of the particular ego that serves as its author must be withdrawn. To that end, the author must remove his cloak of anonymity and explicitly stand present. For, as long as a detached and anonymous ego covertly presides over the text, the Self's Projection will not be retracted in earnest. Then let me introduce myself properly.

As the dreams I have inserted in the first two chapters bear out, this writing comes not from some deus ex machina but from a flesh-and-blood being who has labored long and wrestled much with it: Steven Rosen (fig. 4.3).

Figure 4.3 Steven Rosen.

The messy and turbulent interior struggles of the writer are part of the roiling subtext from which the text bubbles up. So Steven cannot simply be bypassed, either through an idealist rendering of the Self that would transcend him, or through an act of post-structural abstraction that would subtract the individual writer from the writing process. This is surely not to say that a self-signifying alchemical text could be *limited* to the self of the particular author, since what alchemy seeks is the concretely universal Self. But to reach alchemy's goal of fashioning the subtle body for the Self's containment, the body must be made real with the existentiality of *this* body, that of he who now writes. So it seems I must put my body where my words are, enacting in the process a dialectic of finite and infinite bodies, of particular and universal egos, of self and Self. With this in mind, let me inquire further into what it means for me to stand present in the text in an embodied way.

I am here in my study in Vancouver, present at this keyboard, working on this text. At the moment, it is 1:21 p.m. on January 19, 2014. Do I truly and fully stand present with this acknowledgment? Consider the posture that is tacitly assumed in this way of regarding "Steven Rosen." He is viewed as a particular person situated in a particular place at a particular moment in time. Thus he is approached in the "forward" orientation, projected as a finite being. This manner of presenting him is in fact but a *re*-presentation. Steven is signified as an object cast before a subject that itself remains anonymous. For, in this alleged self-signification, Steven actually has divided himself into *subject as object*—a certain person who can be defined in terms of certain objectifiable characteristics—and *subject as subject*—a still elusive anonymity. How then can we say that the author stands fully present here and now? Objectifying himself in this way, the "here" and "now" that he signifies are in truth *elsewhere and otherwhen* relative to a resituated here and now, an originating presence whose locus is the new anonymity established by the split. In rendering Steven an object, the actual here and now is displaced to the nameless subject as *subject*. If the author is to genuinely stand present in the text, to be *here now*, an objectified semblance of him clearly will not suffice. Evidently, he must present himself with the *(w)hole* of his being, both object (whole or plenum) and subject (hole or void); he must be here now with the entirety of his paradoxical Self. It is certainly true that Self-signification could not be realized were the author to write in a purely anonymous fashion. The person behind these words does have to make his presence felt. Only by dropping his veil of anonymity and addressing the question of the transpersonal Self in a more *personal* way can he approach that Self. But a more complete approach to Self requires a fuller presencing than can be provided through self-objectification, and this, in turn, means that the forward thrust by which Steven objectifies himself must be drawn back in.

This is far easier said than done, of course. Although the author of this text may wish to counteract his self-objectification by Proprioceptively accessing his pre-objective infinite bodily source, he is confronted with the compelling sense that the body behind these words is only that of the objectified finite being, Steven Rosen. In typing these abstract words, no doubt I could gain a measure of bodily awareness simply by bringing my attention to my hands and fingers as they touch the keyboard. The proprioceptive sense I would thus obtain of the working of my muscles is readily achievable. But there is one thing I normally do not proprioceive, namely, the *"I" itself*, that core subjectivity which defies objectification. As noted above, the "I" in question is the thinking subject or Cartesian *cogito*. Philosopher Maurice Merleau-Ponty referred to it as "that central vision that joins the scattered visions. . . . that *I think* that must be able to accompany all our experiences."[27] Social psychiatrist Trigant Burrow termed it the *"I"-persona*.

In chapter 2, I mentioned Burrow's call for human beings to gain proprioceptive awareness of the organismic basis of their divisive symbolic activity. Said activity is tied to thinking and language, and Burrow deemed it responsible for the contemporary fragmentation of human society. According to Burrow, it is the "I"-persona that operates behind the scenes to drive thinking and foster division. Through ceaseless acts of objectification, this anonymous thinking subject governs every facet of human experience and behavior in today's world.

The "I" is the prime "identity operator" upon which all my specific operations are based; it is the "master word" that lies behind all these particular words.[28] And this "I think" is what must be Proprioceived in order to surpass Steven's finite particular body and gain access to the generic thinking body. That is, the Self must think itself. For the "I" is none other than the Self in the mode of concealing itself and splitting itself off for the purpose of Projecting finite particular egos to advance its individuation. In the "switching of gears" presently called for, the Projective thrust of the Self is counteracted by a backward movement against its grain that permits its action to be consciously apprehended. This act of Proprioception is what allows the Self to withdraw its Projection of Steven Rosen and stand present as the true author of the text.

For the Proprioception that is called for, the stereoscopic meditation upon the Klein bottle (fig. 4.2) must be carried further. I noted above that merely proprioceiving the eyes as one works stereoscopically with the Kleinian signifier will not suffice for engaging the Self per se. Again, it is not just the eyes that must be attended to but the "I" that lies behind them. What I now suggest is that the movement backward into the "I" culminates in the *brain*. My proposal is based on the work of Burrow.

PROPRIOCEIVING THE BRAIN

According to Burrow, the "I"-persona has a distinct site of operation within the human organism. It is found in what Burrow termed the "cerebro-ocular" region, that is, the cerebral cortex of the brain and its associated organ of vision (the visual cortex is localized in the occipital lobe of the cerebrum).[29] Burrow pointed out that it was through the phylogenetic development of the cerebral cortex that our abstract language and symbolic activity first arose. Therefore, to gain an immediate sense of this activity, it seems one would have to gain access to the cerebrum. But this conclusion was informed by more than a simple logical deduction. Burrow claimed to have had a spontaneous experience of the "I"-persona's bodily base, one that profoundly influenced all his subsequent research. After a prolonged period of interpersonal strife involving the members of the group that he had established to investigate such "I"-based conflict, he began to notice a distinctive pattern of tension around his eyes and forehead. Burrow recognized in this the bodily expression of the "I"-persona.

I am proposing that, to enact the *unio mentalis* in the most concrete way, a Proprioception is required wherein the author of the Kleinian text achieves tangible awareness of the Self's Projection of his basal identity. In Burrow's terms, this entails bringing attention to the ocular-facial or "cephalic segment," that is, to the area of the body around the forehead and eyes associated with the workings of the brain.[30] Burrow would caution us not to confuse the "I"-persona that resides therein with the particular ego of the allegedly isolated individual. He might say that this persona is the *species-wide* "ego" or "subject" that lies behind the appearance of individual subjectivity—the subjectivity of Steven Rosen, for example. While it is through the "I"-persona that the impression is created of merely isolated, disembodied subjects, the generic "I" itself is no disembodied subject. Rather, it is the *bodily process* that is central to human functioning as a whole. Therefore, when Burrow became attentive to the "I"-persona rather than continuing to be unwittingly governed by it, he experienced this palpable pattern of tension around the eyes and forehead against the "tensional pattern of the organism as a whole."[31] He was thus presumably able to apprehend in an immediate way what he called the "solidarity of the species"—what we are calling the Self.[32]

To repeat the basic proposition: the "I"-persona that Projects our sense of disembodied identity is in fact the embodied Self in the mode of Self-concealment and Self-Projection, processes that have served the interest of individuation in the past. For the switching of gears that is presently required to realize individuation in full, the Self's Projection of ego must be consciously withdrawn. And in order for that to happen here and now as a full-fledged *coagulatio*, it must happen through the coagulation of this text. Here the Self-Projection to be retracted via Proprioception is that of author Steven Rosen.

Following his first spontaneous Proprioception of the generic organism, Burrow sought to cultivate the experience in a systematic practice he named "cotention."[33] He described his procedure as one of setting aside daily experimental periods in which he "adhered consistently to relaxing the eyes and to getting the kinesthetic 'feel' of the tensions in and about the eyes and in the cephalic area generally."[34] Burrow's proprioceptive practice was rooted in his conviction that the "I"-persona manifests itself concretely in the muscular activity of the *eyes*, continually engaged as they are in minute acts of binocular convergence that serve to objectify the world. We know that a similar proprioception of the eyes is involved in the act of stereo-viewing the Klein bottle, one that can eventuate in a Proprioception of the "I." The importance of the eyes to the operations of the "I" (the thinking subject or *cogito*) is confirmed by philosopher Drew Leder:

> The "mind's" knowledge of a stable, copresent, external world is . . . largely derived from the eye. Thus, in a model such as Descartes's, which emphasizes the epistemological subject [the *cogito*] and regards truth as involved with definiteness and permanence, visual experience will be attended to above all others. It emerges as "the noblest and most comprehensive of the senses" as Descartes writes at the beginning of *Dioptrics*, his work devoted to vision.[35]

So it seems that, if "I" am to engage in the Proprioception of the "I" serving as master word of this text, "my" practice evidently has to include obtaining a bodily sense of the very organs so central to the act of textual signification: the eyes. And this aim is facilitated by the stereoscopic apprehension of the Klein bottle (fig. 4.2).

To be sure, the Proprioception cannot *stop* at the optical surface of the body. The movement backward must pass *through* the eyes. It must penetrate to the "I" that projects Steven Rosen, to the "I"-persona housed in the cerebral cortex of the brain. Proceeding in reverse, the brain is experienced very differently than when approached in the forward gear. No longer is it taken as a circumscribed lump of matter, a finite particular object from which the observing subject is detached. Proprioceptively apprehended, the brain, as flesh of the "I," is *sub-objective*, a psychophysical modality with a generic, encompassing scope. The brain thus experienced—far from being a discrete organ encased within the cranium of a particular organism—expresses an aspect of what Burrow called the "phyloörganism."[36] The Proprioceived brain is the communal body of cognitive humanity. Alchemically speaking, it is the Self that is realized in this Proprioception of the brain, a process wherein author Steven Rosen's Kleinian text becomes the text of humankind at large.

Recall from chapter 2 von Franz's description of ancient attempts to endow the Self with "a new body," a "'body of a spiritual kind,'" a *subtle body*. I venture to say that this can be facilitated through the Proprioception of the brain enacted with the Kleinian text. We are aware of the *uroboric* nature of the subtle body: it is a body or vessel that twists back into itself in proprioceptive fashion, as in alchemy's *retorta distillatio* (see chapter 2). By now we are also well acquainted with the topological counterpart of alchemy's subtle body—our Klein bottle. What I have not established until the present is that the brain itself is known to possess Kleinian properties. According to mathematical theorist Diego Rapoport, the Klein bottle plays a critical role in the "inner topological and functional structure" of the brain:

> It is most remarkable that the Klein bottle can be found . . . as the solution of [the] representation of topographic maps on the neurocortex. . . . In the visual cortex there is an orderly map of visual space; furthermore, symmetry properties of [cells] . . . lead naturally to the construction of the Klein bottle. So the geometry of visual space has a representation [in] the visual cortex and . . . at the fundamental level of cells, the topology of the Klein bottle is naturally present. Furthermore, the topographic representations are arranged topologically, and most remarkably, there is experimental evidence that supports that these maps can be represented by the Klein bottle.[37]

The Kleinian structure of the brain thus confirms it as a good candidate for the subtle body of the cognitive Self. (In coming chapters, we are going to see that the brain is but one of several organ systems constituting dimensions of the Self.)

To sum up what I am proposing: (1) Burrow's "phyloörganism" is a generic or universal body that can be associated with the Self; (2) the Self's subtle body has the uroboric quality of moving back into itself, as with the alchemical vessel; (3) the same Proprioceptive action is present in the self-reflexive Klein bottle; and (4) the structure of the brain, particularly the visual cortex, is Kleinian in character. All in all, we can say that the Proprioception of the brain intended for this text should indeed be a Kleinian movement that would endow the Self with a subtle body.

KLEINIAN MIRRORING

Presently, I am again bringing my attention to my eyes as they stereo-view the Klein bottle of figure 4.2. I direct my awareness to the place where the Klein bottle intersects itself. If the hole created by the self-intersection is the gateway to the "fourth dimension," if this is where subjectivity enters the picture, and if the subjectivity in question is not some abstraction but is concretized as the viewer's own subjectivity, then the region of self-penetration must function as a kind of *mirror*. Regarding the movement of the Kleinian signifier back into itself not as something happening "out there" in an object that is merely external to me, but as a movement back *in here*, it is as if these eyes of mine that view the self-intersection are viewing themselves—as if, in proprioceptively stereo-viewing the Klein bottle's reentry into itself, I enter the hole this creates to perceive *my own eyes* engaged in the act of viewing. Of course, ordinary mirroring takes place in ordinary space so that, when looking at my eyes in the glass, they appear as objects detached from this viewing subject. The *Kleinian* mirror is different. With its involvement of alchemy's added psychophysical dimension, the mirroring should indeed close the gap between the viewing subject "in here" and the object viewed "out there," since—in the Kleinian vessel—inside and outside flow together as one.

There is a remarkable work of graphic art that seems to map the Kleinian mirroring effect. I am referring to M. C. Escher's *Print Gallery* (fig. 4.4).

Figure 4.4 M. C. Escher's *Print Gallery*. © 2014 The M. C. Escher Company—
The Netherlands. All rights reserved. www.mcescher.com.

In figure 4.4, a boy is shown standing in an art gallery gazing at a print on the wall. In viewing this Escher print itself, which we may call the "master print" (to distinguish it from the prints depicted in the gallery), we observe the boy. He appears to be embedded in the two-dimensional world of the master print, while we experience the three-dimensional field lying outside. However, Escher has provided him with a means of departing from his two-dimensional frame of reference and aligning his consciousness with our own.

Let us review Escher's own commentary on *Print Gallery*:

> We have here an expansion which curves around the empty cen-
> tre in a clockwise direction. We come in through a door on the

lower right to an exhibition gallery where there are prints on stands and walls. First of all we pass a visitor with his hands behind his back and then, in the lower left-hand corner, a young man who is already four times as big. Even his head has already expanded in relation to his hand. He is looking at the last print in a series on the wall and glancing at its details. . . . Then his eye moves further on from left to right, to the ever expanding blocks of houses . . . and this brings us back to where we started our circuit. The boy sees all these things as two-dimensional details of the print that he is studying. If his eye explores the surface further then *he sees himself* as a part of the print.[38]

If we take the liberty of supposing that the row of prints depicted by Escher is a single, moveable print, we may imagine it beginning its circuit at the lower right end of the gallery, rotating clockwise from there, then expanding around the blank spot in the center to take on three-dimensional perspective. In this way, the print manages to "swallow" the master print in uroboric fashion. And when the boy in the picture follows this movement with his eyes, he comes to view himself as we three-dimensional beings view him, as a circumscribed detail of a two-dimensional canvas.

Summing up the transformation enacted in *Print Gallery*: we revolve around a central blank spot or hole, a dimension is added, and consciousness expands recursively so that a figure in the painting transcends his lower-dimensional context to see himself mirrored back. And this is essentially similar to what happens in advancing the work with the Kleinian signifier, where the added dimension is the psychophysical one. At the same time we stereo-view the hole created by the self-intersection of the Klein bottle appearing out in front of us, we must turn proprioceptively backward into the "inner" hole, the "blank spot" or "blind spot" in our visual field. Just as visual experience is blocked where the retina meets the optic nerve due to the absence of receptors there to register the information, so too, phenomenologically, we normally cannot receive information about the *source* of our vision. As Leder put it, ordinarily, we "cannot see our own seeing," since our seeing is what we see *from*.[39] In the customary "from-to" posture, the "from" is a "nullpoint," a mere point of departure for experience, one that itself is *not* experienced.[40] But, in assuming the proprioceptive posture, the usual forward orientation is reversed. Now we are to *arrive* at our "point of departure," *enter into* the previously inconspicuous null point, hole, or "blind spot" in the field of vision. And this retraction of light into the point-origin of the visual field, this backward movement of the eye, opens up a new dimension leading us reflexively to "that central vision that joins the scattered visions. . . . that I *think* that must be able to accompany all our experiences"—what Burrow called the "I"-persona.[41] Proceeding in this manner, the proprioception of the eyes at the optical surface of the body deepens into a Proprioception of the "I" that terminates in the cerebral cortex of the brain. By so plumbing the depths of the hole "in here"— following it into the interior recesses of eye, "I," and brain—we discover its identity with the hole "out there," the hole in the Kleinian "object" that leads to the "fourth dimension." Let me reiterate that the brain in question is not the one known to the Cartesian ego as an object in space encased in a finite body. Instead, it is the brain associated with the Self's infinite Kleinian body, with the body of humanity at large. In thus realizing its subtle body, the Self is poised to assume authorship of the Kleinian text.

THE NIGREDO

Still another time, I am stereo-viewing the Kleinian signifier and attending to its self-intersection as I proprioceive my eyes, with the intention of deepening my proprioception so as to enter the hole leading to the "I." Yet I find myself resisting the process as much as ever. Why such resistance? It is because, at bottom, my very life seems to be at stake. Entering the "black hole" means engaging in the *nigredo*, the alchemical ordeal in which I face my demise as a finite being detached from other human beings and from the rest of nature. But, resist though I might, this death of the mental ego must be endured to complete the *coniunctio* that will bring rebirth. To realize (w)holeness, I must enter the *hole*.

Alchemy associates the *nigredo* with *decapitation*. The head is the seat of the post-Renaissance ego, so it is the head that must be "severed." What we actually have, of course, is a *paradoxical self-decapitation*, one in which the head that is detached also remains intact!

Picture again the uroboric serpent swallowing its tail (fig. 4.5). Bit by bit it ingests itself. Let's assume it can continue this process to its logical conclusion. It seems the snake must then wind up biting its own head. Yet, paradoxically, the head that does the biting stays in one piece! Does the serpent sacrifice itself by devouring its own body? Yes, it does. But— like Schrödinger's cat that is both alive and dead in quantum superposition—the serpent lives and is nourished by its self-sacrifice. In fact, the Proprioceptive sacrifice by which the snake penetrates itself in Kleinian fashion renders its finite body infinite—not as a body with indefinitely extended boundaries but as a subtle body with paradoxical boundaries, boundaries that are not boundaries ("the outside to the inside, the inside to the outside"). In addition, the serpent's body becomes immortal—not as endlessly extended in time but as possessing an aspect of timelessness. The self-decapitation of the uroboros expresses symbolically the alchemical transmutation of the mental ego. With it, the circumscribed brain of the ostensibly separate individual becomes alchemy's subtle brain, the collective brain of humanity.

Figure 4.5 Image of the uroboros adapted from an image appearing in the *Chrysopoeia of Cleopatra*, an ancient Egyptian alchemical text (reprise of fig. 1.1).

But, alas, I am loath to make the sacrifice. For here I am, comfortably ensconced as homunculus within the familiar confines of my seemingly well-bounded skull. It feels like I've dwelled in here forever and it is my home. I'm deeply and profoundly identified with this finite being projected as residing in the center of my head behind my eyes. This "I," this Steven Rosen, appears to be my absolute ground. So how can "I" even think of surrendering it? The content of my words may proclaim the need for the "I" to give up its dominant position, but with every word "I" utter, the "master word" implicitly thrusts itself forward: "I," "I," "I," "I."

Still and all, the paradoxical self-decapitation required does not mean simply obliterating the "I," but Proprioceiving it. So rather than just engaging in an act of self-mutilation in which the "I"'s forward thrust is cut off, what is called for is the backward movement to the source of the Projection.

This is certainly difficult enough. For the movement necessary, I must push back against all my impulses for satisfying closure, back into a void of indeterminacy, a kind of murky purgatory where there is nothing whatsoever to hold on to and everything seems to slip away. It is this experience that drives me away from my work on the text, sending me in search of immediate closure. I write a line, then check my e-mail. I write another line, then look for political news on the Internet. I go back briefly to edit what I've written, then play a game of electronic solitaire. Must I cease and desist from these escapist departures, or is the actual challenge to bring my awareness to them in such a way that, when the impulse to take flight seizes me, I trace it backward to its source rather than just allowing myself to be carried along by it?

I do keep coming back to the task of Proprioception, knowing that it is not something I can get simple closure on, since, by its very nature, it is geared toward counteracting said closure. The Proprioceptive journey is not like getting a rewarding e-mail message or winning a contest. It does not end in a well-defined outcome tied up with a bow. Much less does it conclude in some final state of transcendence that brings salvation once and for all. With Proprioception, the journey *is* the "end"; the ultimate "product" is the process itself.

In light of all this, what can be said of the *unio mentalis*, the first *coniunctio* of the alchemical opus? We can say that this *coincidentia oppositorum* is no ideal state of mystical oneness simply transcendent of ordinary life experience. As a *coagulatio*, the opening conjunction involves a continual movement backward into the body against the ego's disembodying forward thrust. What is paradoxically united in the *unio mentalis* is the forward action of the mental ego or *cogito* and the backward counteraction that brings us to the ego's source in the embodied Self.

Have I been expecting to complete the first conjunction and end the chapter with a flash of lightning, a clap of thunder, and an angelic choir intoning the achievement of Heavenly Rapture? The concrete reality is that the chapter must "end" in the ongoingness of Proprioceptive process. Everything has changed and nothing has changed, as I am poised to get up from this chair and go right out to pick up my car from the Toyota Service Department upon completing these words.

CHAPTER FIVE

Second Stage of Alchemical Conjunction
Cogito and Anima

The *unio mentalis* is "only the first stage of conjunction or individuation," says Jung.[1] We know that beyond this distillation of thinking via
Proprioception of the brain, further distillations are required involving
denser, more concrete *coniunctios*. The present chapter focuses on the
second stage of conjunction, that wherein the embodied mind is brought
into harmony with the realm of the body concerned with feeling. Granting that the Klein bottle, functioning as Hermetic vessel or subtle body,
contains the *unio mentalis* in the opening stage, I submit that two conditions must be met for the containment of alchemical process in stage
two: (1) the Kleinian vessel must be sealed a second time in a denser
medium and (2) a new, more densely embodied vessel must make its
appearance. Of course, these vessels are not just objects in space cast before detached subjects. They are *spatio-sub-objective*: they *merge* object,
space, and subject. The vessels in question can therefore be grasped as
topo-dimensional worlds unto themselves. Before going any further, let
me attempt to clarify the *dimension number* of the Kleinian world.

In foregoing chapters, I noted that the Klein bottle brings into play a
"fourth" dimension. I bracketed the word "fourth" to indicate that the
extra dimension is not just another exterior space extending at right
angles to the first three spatial dimensions; instead it is linked to the
interior dimension of the psyche. Perhaps, rather than characterizing
the Klein bottle as three-dimensional or four-dimensional, it would be
best to describe this spatio-sub-objective vessel as "3 + 1-dimensional,"
with the understanding that the objective three-dimensional spatial
world and the unitary dimension of subjectivity are not just related to
one another in an external, additive fashion, but fully interpenetrate
each other.

Though Jung did not allude to the Klein bottle, it is interesting that he gave archetypal significance to 3 + 1-dimensional structure, regarding it as a "space-time quaternio" constituting "the organizing schema par excellence among the psychic quaternities."[2] The 3-D spatial aspect of the Klein bottle can readily be associated with the spatial aspect of Jung's *quaternio;* this feature clearly aligns with objectivity as well, given the role of space as the fundamental framework for objectification. Does the subjective aspect of the bottle align with time? Evidently, it does. Phenomenological philosophers such as Husserl, Heidegger, and Merleau-Ponty have probed deeply into the intimate relationship between temporal experience and core subjectivity. Merleau-Ponty, for example, begins a chapter on temporality by declaring: "In so far as . . . we have already met time on our way to subjectivity, this is primarily because all our experiences . . . arrange themselves in terms of before and after, because temporality . . . is the form taken by our inner sense, and because it is the most general characteristic of 'psychic facts.'" The subject is essentially temporal "not by reason of some vagary of the human make-up, but by virtue of an inner necessity."[3] On this basis, we may indeed conclude that the unitary subjective component of the 3 + 1-dimensional Klein bottle is temporal in nature, thereby linking it to the temporal aspect of Jung's 3 + 1-dimensional space-time archetype. How does this relate to my claim that the second stage of alchemical conjunction requires a second sealing of the Kleinian vessel, now in a denser medium? What we are going to see is that the "denser medium" involves a more concrete order of space and time.

MYTHIC DIMENSIONALITY AND ANIMAL EMOTION

Although the *cogito* can be said to reach the high point of its development with the *unio mentalis* of stage one, its individuation is in fact not complete without Proprioceptively revisiting its "low points" in subsequent stages of the *coniunctio.* We have seen that immature forms of thinking were pervasive in pre-Renaissance culture, modes of cognition lacking clarity and definition. In previous chapters, I alluded to mythic and magical types of thinking but made no attempt to differentiate them. Cultural philosopher Jean Gebser did lay out their differences and in the present chapter we will focus on the less primal of these, namely, mythic thinking, since this is the sort of cognitive activity that comes into play in the second stage of conjunction.

Gebser too viewed structures of consciousness in dimensional terms.

Whereas post-Renaissance rational thinking is associated with three-dimensional space, the mythic structure of consciousness is for Gebser "the expression of two-dimensional polarity."[4] He relates mythic consciousness to "the circle, the age-old symbol for the soul" and portrays it as "an encompassing ring on a two-dimensional surface":

> It encompasses, balances, and ties together all polarities, as the year, in the course of its perpetual polar cycle of summer and winter, turns back upon itself; as the course of the sun encloses midday and midnight, daylight and darkness; as the orbit of the planets, in their rising and setting, encompasses visible as well as invisible paths and returns unto itself.[5]

Yet, while mythic consciousness "encompasses, balances, and ties together all polarities," the tension between poles is not simply resolved. No third element is introduced here to serve as a "synthesis of opposites" sublating the interplay of the two in a "higher-order oneness." Mythic awareness therefore should not be confused with three-dimensional mental-rational consciousness. On the contrary, the mythic structure is an inherently ambivalent, "irrational" mode of experiencing; it flows to and fro, from one pole to its complement and back again, neither integrating the poles nor denying the validity of either. Gebser makes the observation that this two-dimensional wavelike rhythm of the soul is accompanied by an experience of time characterized by a retrospectiveness, a continual "harking back to what was," a returning again and again to the "beginning," as the ocean waves rise, crest, and fall back down into their troughs, only to rise again. So as to distinguish cyclical/mythic time from the linear "progression into the future" subsequently to develop along with the advent of mental/spatial consciousness, Gebser characterizes the former as "temporicity."[6]

In addition to linking mythic consciousness to the realm of the soul, Gebser variously associates it with the dream state, the element of water, the astral and lunar worlds (stars and moon), movement and kinesis, and with the organ of the heart.[7] I propose that, in functional terms, all of these qualities are related to *feeling* or *emotion*. The relationship between emotion and the heart is a commonplace that I will not elaborate on here, since my immediate purpose is only to note that there *is* a relationship. For now, we can also take for granted the link between emotion and dreams; indeed, it is this that forms the basis of classical psychoanalysis. Most of the other cognates of the mythic given above are touched on by Jung in *Mysterium Coniunctionis*, in a section where he examines the symbolism of the moon in alchemy.

First we have the tripartite connection between the moon, the soul, and water: "The relation of the moon to the soul, much stressed in antiquity, also occurs in alchemy, though with a different nuance. Usually it is said that from the moon comes the dew, but the moon is also the *aqua mirifica* . . . [the] 'Mercurial water' and 'fount of the mother.'"[8] Continuing in a similar vein, Jung associates Luna with "moisture" and with the "'waters under the firmament.'"[9] He also intimates a connection between the moon and the stars: "The moon appears to be in a disadvantageous position compared with the sun. The sun is a concentrated luminary: 'The day is lit by a single sun.' The moon, on the other hand—as if less powerful'—needs the help of the stars."[10] Thus the moon is linked to the *astral* domain. In the esoteric literature, much is made of the "astral plane," and the "astral body." And, typically, astrality is related in these writings to the moon and to the element of water.[11]

Now, a primary characteristic of the astral body is kinesis or motion; the literature is filled with references to "astral travel" and "astral projection." And indeed, this is an essential feature of the *soul*, which Jung generally takes to be synonymous with the archetype of the *anima*. The *anima* is an aspect of the unconscious often symbolized by the snake (fig. 5.1), which bespeaks the "active, animal principle," and is related to "emotionality and the possession of a soul"; Jung links "animation of the body and materialization of the soul."[12] So we have here the correlation of the soul or *anima* with animation or motion, and with *e*-motion. Regarding the latter, Jung says elsewhere that "since the soul animates the body . . . she tends to favor the body and everything bodily, sensuous, and emotional."[13] Also on the subject of emotions, Jung says, "The appetites . . . pertain to the sphere of the moon: they are anger (*ira*) and desire (*libido*). . . . The passions are designated by animals because we have these things in common with them."[14] For his part, Gebser relates "the emotive sphere symbolized and designated by the diaphragm and heart" to mythic consciousness.[15]

From the foregoing we can conclude that mythic thinking is an elemental form of cognition essentially colored by emotion. Sealing the Kleinian vessel a second time means moving awareness Proprioceptively backward into the feeling-toned mythic cognition that prevailed at an earlier time in human history. But mustn't the Kleinian containing vessel possess the same dimensionality as the mythic consciousness it contains? If the dimension number of the Klein bottle is 3 + 1, how then

Figure 5.1 Two images of woman with snake, from the dream art of Jungian analyst and artist Maria Taveras.

could mythic cognition be *two*-dimensional as Gebser claims? What we are going to see is that, while mythic cognition indeed is closely affiliated with a sphere of emotion possessing lower dimensionality, as a form of *thinking* it is itself in fact 3 + 1-dimensional, though only incipiently so. To begin to get a handle on this, let us turn to the question of phylogeny.

Jung's association of the soul or *anima* with the "animal principle" and his remark that the "passions" are what we have in common with animals point significantly to phylogenetic implications which, for the most part, are neglected by Gebser, whose focus is primarily on the evolution of *human* consciousness. In this regard, the old idea of the "animal soul" is noteworthy. Philosopher Antony Flew defines this as an "analogue in animals of the human soul or mind," one that suggests a principle of animation in the world existing independently of the human sphere of action.[16] A more general expression of the "animal soul" is the "world soul" or *anima mundi*, a concept "founded on the view that the [nonhuman] world is productive of life and animation, and can therefore be regarded as itself animate."[17] In fact, Jung makes frequent reference to the *anima mundi*, demonstrating that its liberation is a central goal of alchemy.[18] I venture to suggest that the imprisonment and subsequent freeing of the *anima mundi* portrayed in the alchemical literature metaphorically expresses the phylogenetic aspect of the developmental drama. The "Spirit" alluded to in the quintessential tale of the "Spirit in the Bottle" discussed in chapter 2 is none other than the *anima mundi*. Though we shall discover that this *anima* is not itself cognitive and is of lower dimension than the *cogito*, by Proprioceiving the 3 + 1-dimensional mythic *cogito*, the *anima* is invited to make its presence felt. (The present chapter deals chiefly with the *animal* aspect of the *anima mundi*. In succeeding chapters, other aspects will be explored.)

Enacting the second *coniunctio* means moving awareness backward into the pre-rational mythic past. To facilitate this, we go back to the writing of the Jungian theorist Erich Neumann. In the *Origins and History of Consciousness*, Neumann states that "primitive man experiences an 'animated' world, while modern man knows only an 'abstract' one. Pure existence in the unconscious, which primitive man shares with the animal, is indeed nonhuman and prehuman."[19] Neumann asserts that, "the dawn man lives his affects and emotions to the full."[20] The transpersonal philosopher Michael Washburn, taking his cue from Neumann, notes that these primordial emotions (today still experienced in human infancy) include a feeling of "exuberance, or overflowing well-being," "waves of bliss," and a sense of "delight" in a world that is enchanted with "an aura of the miraculous."[21] However, Neumann points out that primal emotionality also includes "instinctive reaction[s]" involving "flight or attack . . . rage, paralysis, etc."[22] He contends that human consciousness has evolved "from the primitive emotional man to the modern man" in a manner that "protects—or endeavors to protect—him from this [excess] of primitive emotionality."[23] The protection is needed because "primitive man . . . like the child . . . was forced into total reaction by any and every content that emerged, and, overpowered by his emotionality and the underlying images, acted as a totality, but without freedom."[24] This "total reactivity of primitive man is no subject for romanticism."[25] Nevertheless, given "the present crisis of modern man, whose overaccentuation of the conscious, cortical side of himself has led to excessive repression and dissociation of the unconscious . . . it has become necessary for him to 'link back' with the medullary region" of the brain, or, what amounts to the same thing, with the "heart-soul which animates the body."[26] I suggest that this "linking back" to primal feeling is what happens in the Proprioception attendant to alchemy's second conjunction. Neumann's alchemical rendering of the process is noteworthy:

As in alchemy the initial hermaphroditic state of the *prima materia* is sublimated through successive transformations until it reaches the final, and once more hermaphroditic, state of the philosopher's stone, so the path of individuation leads through successive transformations to a higher synthesis of ego, consciousness and the unconscious. While in the beginning the ego germ lay in the embrace of the hermaphroditic uroboros, at the end the self proves to be the golden core of a sublimated uroboros, combining in itself masculine and feminine, conscious and unconscious elements, a unity in which the ego does not perish but experiences itself, in the self, as the uniting symbol.[27]

Now, Neumann places strong emphasis on the overriding importance of the *animal* in primordial human affairs. He notes that "the animal forms of the gods and ancestors originally symbolized and expressed man's oneness with nature."[28] In native bands of the Pacific Northwest, for example, the initiation rites crucial to the entire course of a person's development centered on

induction into the spiritual world of the ancestral totem . . . [which] is very often an animal. . . . The totem is an ancestor, but more in the sense of a spiritual founder than a progenitor. Primarily he is a *numinosum*, a transpersonal, spiritual being. He is transpersonal because, although an animal, a plant, or whatever else, he is such not as an individual entity, not as a person, but as an idea, a species.[29]

Through totemic initiation, the individual acquires a "guardian spirit": "This spirit, who may be lodged in an animal or a thing, introduces into the life of the initiate who experiences him a whole sequence of ritual obligations and observances, and plays a decisive role among all shamans, priests, and prophetic figures in primitive societies."[30] Later Neumann comments on how the expansion of human consciousness via "secondary personalization leads finally to the local deities becoming heroes and the totem animals domestic spirits. . . . In consequence, the human and personal sphere is enriched at the expense of the extrahuman and transpersonal."[31] Taking the history of ancient Egypt as an example, Neumann documents "the increasing humanization of the gods." He shows how, from the First Dynasty to the Third, the animal figures that predominated were gradually transmuted into human ones, so that, "from the Third Dynasty onwards the human form becomes the rule. The gods establish themselves in human form as lords of heaven, and the animals retire"[32] (fig. 5.2).

Figure 5.2 The Egyptian goddess Hathor, shown in earlier form as a divine cow (left) and in later form as a woman (right). (Courtesy of Jeff Dahl, Wikimedia.org)

In *Coming to Our Senses*, Morris Berman makes clear the extent to which modern industrial society has lost touch with its animal heritage. According to Berman, the Paleolithic human being saw the animal as an Other, but one with whom he or she was intimately identified. In fact, the earliest humans were so deeply immersed in the animal world that it might be plausible to say that, in a sense, they were more animal-like than human. "Animal life was everywhere, even in the skies," says Berman, and "animal movement, the animal body, was the model of human expression in hunter-gatherer society."[33] It is not surprising then that "the first art, which can be seen in the caves of Altamira and Lascaux, is about animal subjects rather than human ones"[34] (see fig. 5.3). Berman goes on to say of hunter-gatherer culture that "human life . . . had no special significance apart from the animal world."[35] And even in the next historical epoch, with the advent of agriculture and the domestication of animals, their profound influence on the human psyche was maintained:

Figure 5.3 Paleolithic cave painting in the Great Hall of the Bulls at Lascaux. (Courtesy of Prof. Saxx, Wikimedia.org)

"Animal cults and symbolism continue into the Neolithic period, still strongly coloring human processes of cognitive and psychological development."[36]

However, the agricultural revolution also brought with it a momentous change: the animal realm was presently divided into the "Wild" and the "Tame." Animals in the latter category, pressed into the service of human needs, were stripped of their otherness, whereas those in the former category were now sometimes regarded more as alien beings to be hated and feared—as *merely* other, rather than as *intimately* so. According to Berman, this splitting of the human and animal worlds constituted the basis for all subsequent dualisms attendant to human affairs. In the earlier hunter-gatherer society, "if there was no sharp divide between Wild and Tame, or Self and Other, there was also no such divide between sacred and profane, or heaven and earth. . . . Domestication changed all this. The fundamental categories that presented themselves were now two—Wild and Tame—and eventually all forms of thought, down to the present day, came to be based on this model."[37]

But I do think we need to draw a distinction between the "dualism" that took hold in the Neolithic Age, on the one hand, and that known to us today, on the other. It was *mythic* consciousness that prevailed in Neolithic times, not the mental-rational consciousness of modernity. And we have learned from Gebser that mythic perception is not character-

ized by dualistic splits and stark oppositions but by a flowing from pole
to pole in which opposing terms retain an internal connection. We may
apply this to the Neolithic relationship to animals. It is true that, in the
Neolithic Age, a distinction came to be made between wild and tame animals; concomitantly, human beings were now somewhat less likely to
experience wild animals in a profoundly intimate way. These correlated
developments are consistent with the general idea that when consciousness begins to divide the world, it becomes more divided *from* that world.
But the Neolithic distinction between wild and tame animals was surely
not as categorical as it would become in modernity, and—despite losing
the deep and sustained Paleolithic intimacy with the animal Other—intimacy was by no means lost completely in the Neolithic. This is borne out
by the fact that, as Berman himself notes, "animal cults and symbolism
[did] continue into the Neolithic period, still strongly coloring human
processes of cognitive and psychological development." The worship of
animals cannot be practiced without some sense of intimacy with them.

While Berman does not bring out explicitly the ambiguous character of the wild-tame distinction in the mythic culture of the New Stone
Age, he does elucidate the extreme manner in which wild and tame become differentiated in modernity. On his account, with the transition to
modern industrial society the very awareness of the animal as wild, as
a fear-inspiring Other, is eclipsed: "fear of the animal Other acquires a
whole new dimension during the modern period; it becomes vague, un-
specified, repressed, and all the more terrifying as a result."[38, D1] Thus, in
modern culture, "Nonhuman Otherness is not merely degraded . . . but
absent."[39] Berman concludes that "fear of organic life and the existence
of the Tame/Wild distinction is so central and pervasive a feature of
modern technological societies that it is, paradoxically, almost invisible.
Like . . . the mind/body split [which derives from it], it is virtually everywhere, so it seems to be nowhere."[40] Zoos, pets, and Disney characters
certainly abound in today's world, but all these give us are "humanized
form[s] of ourselves, not a true Other."[41]

Significant for our purposes is the potent *emotionality* of the animal Other. Berman notes that "Paleolithic men and women took their
cues from body feelings and the movements of animals. This was a life
governed by shifting moods rather than the demands of the ego."[42] The
contemporary social phenomenon of psychotherapy alone makes it more
than clear that *modern* emotional life tends to be "bottled up." When the
"cork" is in place, emotions are tame, well socialized, controlled by cognition. Of course, the cork does tend to "pop" from time to time, where-

upon emotions rush out regressively and we are flooded by them, as the youth of the Grimms' fairy tale experienced. Only with the second sealing of the cognitive Kleinian container can the "Spirit Mercurius"—that is, genuine animal emotion—emerge in a full-fledged and constructive fashion. In topo-dimensional terms, what emerges from imprisonment in the 3 + 1-dimensional Klein bottle and enters into harmony with it is a 2 + 1-dimensional body possessing a topology of its own, as we will soon see. While Berman says nothing on the matter of dimensionality, confirmation of the two-dimensional character of the animal's emotional space comes from two independent sources.

Comparing human consciousness to that of animals, the nineteenth-century philosopher Jules Lachelier concludes that animals "are provided with the same senses as we, but it is probable that these senses *move* them much more than they teach them and that these impressions themselves are entirely subordinated to their organic feelings."[43] Lachelier associates animal being or the "animal *self*" with "desire" and "purpose," with the "will to live," which he views as the "fundamental emotional state."[44] This emotional order of consciousness is seen as played out in "two-dimensional space or surface."[45] By contrast, human consciousness involves a higher order of reflection in which "we project outside of ourselves solid objects by adding to the two dimensions of visible space that which is only the imaged affirmation of existence—depth. . . . Three-dimensional space, individual reflection and reason—these are the elements of a . . . consciousness which we have . . . called intellectual."[46]

The other confirmation of the two-dimensionality of animal emotion comes from the Russian theosophical philosopher P. D. Ouspensky. Ouspensky also saw the animal world as a world of emotion: "In reality the animal does not reason its actions, but lives by its emotions, subject to that emotion which happens to be strongest. . . . Its actions are directed not by thoughts but principally by emotional memory and motor perceptions. . . . Any perception of an animal, any recollected image, is bound up with some emotional sensation or emotional remembrance—there are no non-emotional, cold thoughts in the animal soul."[47] Ouspensky then asks: "How does the world appear to the animal?"[48] His answer:

> The world appears to it as a series of complicated moving surfaces. The animal lives *in a world of two dimensions*. Its universe has for it the properties and appearance of *a surface*. And upon this surface transpire an enormous number of different movements of a most fantastic character.

Why should the world appear to the animal as a surface? First of all, because it appears as a surface to *us*. But we *know* that the world is not a surface, and the animal cannot know it. It accepts everything just as it appears. It is powerless to correct the testimony of its eyes—or it cannot do so to the extent that we do.

We are able to measure in three mutually independent directions. . . . The animal can measure simultaneously in two directions only—it can never measure in three directions at once. This is due to the fact that, not possessing concepts, it is unable to retain in the mind the idea of the first two directions, for measuring the third.[49]

What I would add to the idea that animals dwell in a two-dimensional space is that they also possess a sense of time, so that animal dimensionality is in fact 2 + 1. The temporality in question is certainly not the linear progression from the past into future characteristic of post-Renaissance human experience. The nature of animal temporality can be inferred from Gebser's description of the temporal experience of the mythic human being who was strongly influenced by the animal. As noted above, Gebser referred to mythic time as "temporicity"—a continual harking back to what was, a returning again and again to the beginning, as the ocean waves rise, crest, and fall back down into their troughs, only to rise again. I suggest that this cyclical time sense, attuned as it is to the seasons and rhythms of nature, is closely akin to the animal's sense of time. This is not to say that mythic humanity's time sense was *identical* to the animal's. The mythic person, as a 3 + 1-dimensional human being, possessed cognitive capabilities and the potential for linear temporal experience, even though these were weakly developed and largely overshadowed by 2 + 1-dimensional animal emotionality and temporicity.

We must of course go more deeply into mythic experience than we have. In order to seal the Kleinian vessel a second time, just writing *about* the mythic will not suffice. With the first sealing of the vessel enacted in chapter 4, mental individuation (the *unio mentalis*) reached its initial climax through an act of embodied self-signification. Jung states that "the second stage of conjunction" involves a "re-uniting of the *unio mentalis* with the body."[50] However, since we have found that the *unio mentalis* is in fact already embodied in its own way, what is now actually required is reunion with a *denser sphere* of the body. In the present chapter, this *coagulatio* is to be realized by an act of *mythic* self-signification. But before turning to that, we must take up a crucial topological question deferred until now. If the cognitive Self is a 3 + 1-dimensional being whose individuation is contained by the Klein bottle, how may we describe the topological vessel that contains the individuation of the 2 + 1-dimensional emotional Self of the animal?

LOWER-DIMENSIONAL TOPOLOGY

The topological investigation conducted in chapter 3 brought out the fact that the Klein bottle does not simply stand alone but is a member of a bisection series: cutting the one-sided Klein bottle down the middle yields a pair of oppositely oriented Moebius strips; dividing the one-sided Moebius strip in like manner produces a two-sided structure called a lemniscate. While the bisection series clearly constitutes a topological family of closely related structures (fig. 5.4), what is not at all evident from taking these structures as mere objects appearing before us in our familiar three-dimensional space is that each one actually comprises a spatio-sub-objective dimension unto itself.

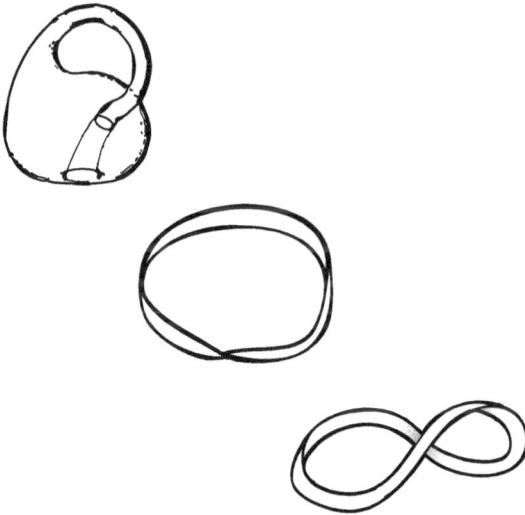

Figure 5.4 Topological family.
From top to bottom: Klein bottle, Moebius strip, lemniscate.

We already know that this is true of the Klein bottle, when the bottle is understood alchemically rather than conventionally. The bottle's self-intersection proves to be the critical factor. In penetrating itself, a hole is created that marks its necessary incompleteness as an object in three-dimensional space. Conventional and alchemical mathematics alike conclude that an added dimension is needed to complete the Klein bottle properly, but while convention deems this extra dimension as just another exterior space in which the bottle is embedded as object, alchemy views it as an *interior* dimension, the dimension of psyche or subject that *merges* with object (see chapters 3 and 4). The self-penetrating, self-containing Klein bottle, in its blending of object and subject, space and time, thus constitutes the uroboric embodiment of the 3 + 1-dimensional Self.

Now, in considering the *sub*-Kleinian members of the bisection series, it seems clear that they *can* be properly completed as objects in three-dimensional space. It is for this reason that, although a paradoxical structure like the Moebius strip can well *symbolize* the integration of subject and object in the 3 + 1-dimensional sphere, it cannot effectively *embody* said integration. That is precisely why we saw the need for dimensional enhancement via the Klein bottle in the previous chapter. The proposition I now offer is that even though the sub-Kleinian members of our topological family indeed cannot embody the dialectic of the 3 + 1-dimensional Self, they do provide embodiments of *lower-dimensional* Selves.

The standard mathematical interpretation of the bisection series says that each of its members is a two-dimensional object (i.e., a surface) embedded in a higher-dimensional spatial continuum (the Klein bottle is also regarded as a surface here, rather than as a solid). What I am proposing from my contrasting alchemical perspective is that *none* of the members of the series actually possesses this status. Since the necessary incompleteness of the Klein bottle in three-dimensional space is plain to us, it is relatively easy for us to see how this structure could defy conventional objectification. But the sub-Kleinian counterparts of the bottle evidence no such incompleteness. Why can we not see Klein-bottle-like holes in these other members of the series? It is because our seeing is three-dimensional. Within this frame of observation, the lower-dimensional dialectics of hole and whole are imperceptible. The principle is implicit in the thought experiment suggested by Rudolf Rucker, discussed in chapter 3.

To reiterate Rucker's operative point, a form that penetrates itself in a given number of dimensions can be produced without cutting a hole in it if an added dimension is available. Rucker illustrates this by having us imagine a species of "flatlanders," dwellers in a world of two spatial dimensions. If these creatures sought to assemble a Moebius strip in their dimensionally limited environment, they would be confronted with the same problem we three-dimensional beings face in constructing the Klein bottle: the "flatlanders" would have to cut a hole in the Moebius (fig. 3.8), for the Moebius would intersect itself in the two-dimensional world, just as the Klein bottle does in our world.

My proposition is that "flatland" is not merely hypothetical but that there really does exist a two- or 2 + 1-dimensional world (though not one inhabited by mathematicians contemplating topological puzzles!). In this lower-dimensional realm, the Moebius plays a role similar to that of the Klein bottle in the 3 + 1-dimensional sphere. Alchemically grasped, the Moebius is a psychophysical body of paradox, a sub-objective, spatiotemporal dimension of the Self. The Moebial order of psyche or subjectivity is not one of human cognition but of animal emotion; in its individuated form, it is an "animal soul" or animal Self of the kind that Neumann alluded to in speaking of the Pacific Northwest totem as a "transpersonal, spiritual being."[51] And Moebial dimensionality is not constituted by linear time externally adjoined to Cartesian space but by a denser, earthier, intertwined world of natural rhythm and circular temporal flow.

The historical accounts of Neumann and Berman previously cited make it clear that, in moving out of mythic culture and into the Greek and post-Renaissance epochs dominated by human rationality, the once-intimate connections with the animal realm have become obscured. Again, in topo-dimensional terms, the 2 + 1-dimensional Moebial world has become "bottled up," repressively enclosed within the 3 + 1-dimensional container. For the Moebial *anima* to be liberated, its higher-dimensional container must first come to be recognized as Kleinian in nature. This Kleinian self-reflection takes place in the conjunction initiated in the last chapter: the Hermetic sealing of the uroboric Klein bottle via the *unio mentalis*. But while the *unio mentalis* is a necessary condition for freeing the *anima*, it is not sufficient. There must now be a *unio emotionalis* that seals the bottle a second time. As noted above, this more concrete closure of the vessel is to be carried out by moving awareness Proprioceptively backward into the feeling-laden cognition of that bygone mythic epoch when the "animal soul" had held sway. It is through resealing the paradoxical Klein bottle in this way that the Moebial *anima* can be freed from it—not simply to become detached from the Kleinian *cogito* but to enter into harmony with it, a harmony of dimensional spheres. In order to enact the required Proprioception in full concreteness right here in this text, an *emotional self-signification* is presently called for.

SELF-SIGNIFICATION OF THE MYTHIC CHILD

In the previous chapter, we sought to carry out a self-signification of the alchemical text. There I noted that, to reach alchemy's goal of fashioning the subtle body for the Self's containment, the body of the text must first be fleshed out. This was done by raising the dimension of signification from these skeletal one-dimensional typographic marks to the stereoscopically realized 3-D Kleinian signifier (fig. 4.2), and by bringing into play the existential life of the one who engages in signification: Steven Rosen, the author of the text. Here I had to drop my veil of authorial anonymity and make my presence explicitly felt. Following the act of standing present in the text, the Self's Projection of my particular self was to be withdrawn. Toward that end, I attempted an alchemical meditation on the Kleinian signifier (née Hermetic vessel) that involves moving backward through the "I" into the generic brain of humanity at large. With this Proprioception, the Kleinian Self as such assumes authorship of the text.

For the self-signification presently required, it is not enough for the adult to stand present. A long-repressed aspect of identity must also make its presence felt. Just as the emotive Moebial Self is repressively contained within the cognitive Kleinian Self, so too a more concretely functioning personage is concealed by the abstractions of the adult. Whereas the latter is given to discursive thinking and rational discourse, the child is governed by mythic mentation and is strongly affected by the tug of the heart. It is as Neumann indicated in his own statement of the widely held idea that ontogeny recapitulates phylogeny: the mythic past of our species "is re-experienced in every early childhood, and the child's personal experience of this pre-ego stage retraces the old track trodden by humanity."[52] For the further coagulation of this text, it is incumbent on its author to retrace Proprioceptively the track he traveled in his own childhood. Let us call the child "Stevie" (the name my parents gave me in my early years). This youth must now step out of the shadows and stand present (fig. 5.5).

Dream of July 3, 2004:

> I'm with my Dad. We're waiting for my mother to come home. I haven't eaten yet and it's late. Has my father already eaten? I think so, but I haven't. I'm doing something, involved with something while I wait.
> Finally, at around 11:30 p.m., a cab pulls up.

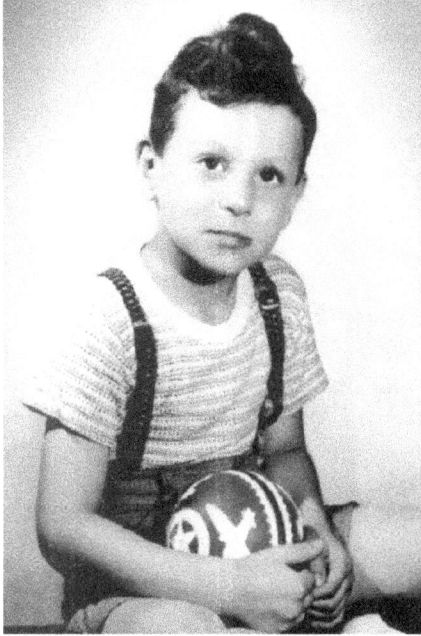

Figure 5.5 Stevie.

"Where have you been?" I ask.

"I've been quite busy with important things," is her answer.

"Am I not important?"

"Yes, but there's been so much to do. I've been out, doing what I need to do."

"Okay . . . actually, I've been busy too. I was playing something and was able to occupy myself until you got back. . ."

I waited so long for her. Can I still have supper at 11:30 p.m.? But it wasn't as if I'd starved myself and was waiting with bated breath. I was busy with other things. But I was waiting nonetheless, and here she was.

The child tries to resign himself to his mother's absence, to convince himself he can go it alone. She has wounded him deeply, but he doesn't want to show it. The little boy has his pride.

Dream of October 9, 2004:

My son David is a child. He has become very angry with me because he's scraped his knee and feels I haven't cared for it properly. I'm a little surprised to hear this. I didn't realize that he thinks I've neglected him. He'd wanted me to put a bandage on his knee. At one point, when he isn't

*there, I look at a small tin box that has his things in it. It's a sad little box
containing a few scraggly items, including some frayed bandages.*

*David is really mad at me. His feelings are badly hurt because I've
neglected him. Why didn't I put a bandage on his knee? I was careless and
impatient in dealing with it. Maybe I quickly rinsed the knee with some
water but hadn't really cared for it; I'd just rushed on to other things. I was
in a "big hurry" and hadn't had time for him.*

*"You hurt my feelings," he says accusingly. His lower lip is trembling.
I'd been insensitive to a little boy's bruised feelings.*

*At some point it seems that David wants his own phone. To comply
with his wishes, I'm having another phone set up but it is right next to
an existing one. I hear the operator's voice loud and clear, establishing
that there is this new connection. And I'm thinking that we don't need it,
that it's a terrible waste of money and a bad mistake. But I don't want to
further hurt David's feelings.*

Little David seems so bereft to me. I find his feeble protests so poignant. Is it only David I've neglected, or the child within *me* as well?

A "new connection" is made, a new voice given. I have some doubts
about this. Is it a mistake to let the child speak out on his own? Should I
be encouraging his appearance in the text? Is this really necessary? I do
think it is.

Dream of March 17, 2005:

*I'm traveling with my son and we're passing through a clearance area
or checkpoint (at an airport, I believe). It's our turn in line. The woman in
charge sends David through. I'm now at the front of the queue and the
people behind me are surging forward a bit, so that the line is threatening
to break, but I contain it. Yet the woman in charge blames me for this
perceived infraction. She singles me out to be left behind. She is chiding
and dismissive. I plead with her, trying every which way to gain her
approval for being allowed to go forward. She does not relent.*

*So I'm sitting near the counter, watching the woman as she permits
everyone else to pass through. I wait and wait, with a deepening sense of
sad resignation.*

*"Oh, won't you let me go through?" No, she won't. So I wait some more,
wait and wait.*

*The child is waiting to be forgiven. Powerless, he waits for her approval.
She is taking her own sweet time about it. Busy with others and having*

put him in his place, she doesn't even think about him. He got what he
deserved and isn't worth any more of her attention.

There's a feeling of impotence in this dream, of being a helpless child.
The child has become invisible to the woman in charge, and he's content
to sit on the sideline, resting in his despair.

Dream of March 8, 2006:

I help a little boy get across a body of water, maybe from Staten Island
to Brooklyn, or to Manhattan. I do this by deceiving the officials who must
approve his passage. Though my motivation is good, what I've done is
illegal.

I find myself dealing with a woman who reminds me of the matronly
and amiable secretary of my old college department—except that this
woman is not so amiable. And she is in charge of things, so I must answer
to her.

At first she doesn't know that I've done anything illegal on behalf of
the boy. But then I'm again asked to help with someone else who is to
make this crossing. Apparently, I'm the one who is putting himself forward
(though others are involved in this illegal act), making myself vulnerable
to discovery.

The woman in charge now begins to suspect me. She begins taking
photographs of me with a strange camera (rectangular, long, and narrow).
She believes I've committed bribery. That's not true! Bribery was not
involved. My motives were benevolent. But quite unjustly, she's convinced
that I'm guilty of the crime.

With each photo she takes, I feel a sense of being exposed, caught in a
trap, unable to defend myself. It's as if, each time she snaps the shutter she
is saying: "I've got you! You're now in deep trouble, and I have complete
power over you."

I now start becoming emotional. I'm pleading with her and begin
to weep: "You think I was involved in bribery? *I would* never *do that,*
never." I'm crying bitterly, and say again, "I would never *do that." And*
she's laughing sadistically. Now she's got me where she wants me, and
is enjoying the sight of me squirming pathetically before her. I feel
humiliated.

Through Steven's dreamscapes, Stevie's presence is sensed. Here the
feelings of the child express themselves: his longing; his anger and hurt;

his sense of guilt, helplessness, and humiliation; his sad resignation. Still, the descriptive containment of these feelings given by the dream stories does not constitute a *Hermetic* containment; the disclosure of the emotional subtext does not add up to a *Proprioception* of the text. For the latter, it will not be enough to gain cognizance of Stevie's particular feelings. *Stevie himself* must be taken back in. The Projection of Stevie as a particular being whose particular emotions are contained in his particular body must be withdrawn. Thus moving backward, the feeling-saturated cognitions of the child come to be recognized as the *generic* cognition of mythic humanity. This second sealing of the 3 + 1-dimensional Kleinian vessel will contain the immature *cogito* so as to bring a greater degree of maturity than had been achieved in the first sealing. Here further individuation will be realized by bringing the mythic *cogito* to light. But something more will happen with the Proprioception of Stevie. The "genie will be let out of the bottle." The noncognitive lower-dimensional *anima* will be released from cognitive containment to enact a Proprioception of its own, one that corresponds to the initial Hermetic sealing of the 2 + 1-dimensional Moebial vessel. In Proprioceptive synchrony, Kleinian and Moebial Selves will then carry out their joint Self-significations.

Before the Proprioception of Stevie can begin, Steven's written text must be distilled to the denser subtextual medium of Stevie's concrete images and spoken words. The child's images are not sharply circumscribed figures cast before the perspective of a Cartesian viewer who regards them with detachment. David Lavery helps us to appreciate the distinction.[53] He draws the contrast between "focal vision" and "peripheral vision."[54] The former gained supremacy in the aftermath of the Renaissance with the rise of perspective (see chapter 1). It was because the Cartesian observer detached himself from the world that he was able to "put things in perspective." Thus standing aloof, he could bring the objects under his visual scrutiny into sharp focus by the process of binocular convergence upon them. Lavery intimates that the eyes worked differently prior to the Renaissance. The pre-perspectival person, participating more fully in the world, being immersed in it, did not primarily focus upon objects held off at a distance. Rather, the sense of being *encompassed* by one's environment led one to experience things and people in a "peripheral" fashion, that is, as being *all around one*. In formulating his account of this earlier way of seeing, Lavery draws from the writings of Owen Barfield, who is subsequently quoted as follows: The "'earlier kind of knowledge . . . was at once more universal and less clear. We still have something of this older relation to nature when we are asleep, and it throws up the suprarational wisdom which many psychoanalysts detect in dreams.'"[55] The old peripheral vision is *oneiric* in character: it possesses a dreamlike quality, as Gebser said of mythic imagination.[56] And the visual imagery of the contemporary child, recapitulating phylogeny as it does, possesses this same quality.

No doubt Steven's dreamscapes are haunted by Stevie, and the dream texts can be distilled by transmuting the adult's written words into the child's mythic images and voicings. To begin, we can picture Steven's son, who appeared as a child in the dream of October 9, 2004. David's *image* (fig. 5.6, left) is presently the text as much as these words I am typing, and were Stevie to stand present to gaze upon the photo, he would view it diffusely. Since he would not bring David's face into sharp relief, this visage could fluidly morph into that of another (fig. 5.6, right), much as in a dream. It is not a *written* word that primarily accompanies such peripheral imaging but one that is *spoken*. When Steven's dream character of March 8, 2006 protests the charge of bribery by tearfully declaring, "I would *never* do that, *never*"—these are in fact *Stevie's* words, and are best expressed not by inscription but with the child's plaintive intonation.

Figure 5.6 David (left) and Stevie (right).

The transition from the child to the adult, or, in terms of cultural evolution, from the mythic to the rational, in fact is marked by the passage to the written word from a word that was spoken. The change is brought out in Gebser's portrayal of how mythic consciousness, characterized by "utterance" and "voicing," gave way to rational consciousness, where "seeing and measuring" were more prominent (the written word is seen, of course, not voiced and heard).[57] In *Interfaces of the Word*, communications theorist Walter Ong attributes great significance to this transformation of the means by which humans communicate.[58] Ong hypothesizes that, over the past 5,000 years, consciousness has evolved from being governed primarily by aural perception, sensitivity to sound, to being controlled by vision. The cultural concomitant of this has been the transition from orality to script, from social transactions based on the spoken word, to those dependent on writing.

Following the inception of script around 3500 BCE, says Ong, "the world of primary orality was torn to pieces by writing and print, which then created, agonizingly, a new kind of noetic and a new kind of culture based on analysis and self-conscious unification."[59] What, specifically, was the primary experience like? Ong offers the following proposition:

> The psyche in a culture innocent of writing knows by a kind of empathetic identification of knower and known, in which the object of knowledge and the total being of the knower enter into a kind of fusion, in a way which literate cultures would typically find unsatisfyingly vague and garbled and somehow too intense and participatory. To personalities shaped by literacy, oral folk often appear curiously unprogrammed, not set off against their physical environment, given simply to soaking up existence, unresponsive to abstract demands such as a "job" that entails commitment to routines organized in accordance with abstract clock time (as against human, or lived, "felt," duration).[60]

Of the spoken word, Ong says:

> [It] is of its very nature a sound, tied to the movement of life itself in the flow of time. Sound exists only when it is going out of existence: in uttering the word "existence," by the time I get to the "-tence," the "exis-" is gone and has to be gone. A spoken word, even when it refers to a statically modeled "thing," is itself never a thing. . . . No real word can be present all at once as the letters in a written "word" are. The real word, the spoken word, is always an event . . . an action, an ongoing part of ongoing existence.
>
> Oral utterance thus encourages a sense of continuity with life, a sense of participation, because it is itself participatory. Writing and print, despite their intrinsic value, have obscured the nature of the word . . . for they have sequestered the essentially participatory word . . . from its natural habitat, sound, and assimilated it to a mark on a surface, where a real word cannot exist at all.[61]

Despite Ong's overdrawn contrast between the "real word" and the written word, and his nostalgic preference for the former, he brings home some important distinctions between these modes of interaction. Of course, while the written word has gained supremacy over the spoken word, we still do speak to each other. Nevertheless, spoken language, though still being transmitted through the concrete medium of sound, has taken on, in its abstract pattern of organization, the character of written discourse. We have come to speak as though we were writing. The clue Ong provides as to the nature of *original* speech—that it is "tied to the movement of life itself in the flow of time"—is reminiscent of what Gebser says of mythic consciousness in general: it is "oceanic," as exemplified in the circular rhythm of the fragment from Heraclitus: "'For souls it is death to become water; for water it is death to become earth. But from the earth comes water and from water, soul.'"[62] How different this is from the prosaic linearity of rational expression later to gain sway. For the text of Heraclitus to be properly conveyed, it must be recited, voiced aloud in the rhythmic tempo of a song.

The Jungian analyst Mary Lynn Kittelson provides further insight into the nature of original hearing and speech in her essay, "The Acoustic Vessel." She begins by taking note of the *repression* of the old sonority that has occurred in our society: "Mostly . . . we work visually. Our words are heard primarily as content. We pay scant heed (consciously, at any rate) to how things sound."[63] Thus:

> Through the [domination of the] eye, the vibratory and participatory aspects of experience fall into the shadows. To be "ear-minded" is to be resonant, layered, slower, sensing things out before the light. . . . Unlike light, whose vibratory nature is a less immediate experience, sound and silence reverberate in a palpable way. . . .
>
> Our collective experience in modern society has supported auditory inattentiveness and misuse of sound. . . . Chatter, complaint, jargon, interminable "how-to's," and the hyped-up headline style of broadcast news are . . . incessant. . . . As adults we have lost our ear-minded center, which was vibrant in previous generations and remains so in much of infant and animal life. According to one neurolinguistic programming study, "Most people in the U.S. do not actually hear the sequence of words and the intonation patterns of what they, or other people, say. They

are only aware of the pictures, feelings and internal dialogue that
they have in response to what they hear."[64]

Using terms like *song, melody, prelude, overture,* and *opera,* Kittelson
emphasizes, both metaphorically and literally, the rhythmic musicality
of uninhibited sonance.[65] *Imagination* surely plays a role in such soundings, as Gebser says of mythic "utterances" and "voicings." The images
may be concretely visual, or acoustical, "as in [the images of] a poem
or song."[66] Like Gebser, Kittelson relates primal sonority to myth in an
explicit way. After observing that soundings of this kind are "noticeably
present in many creation myths," she focuses on the Greek myth of Echo,
which voices "longing, dreams, loneliness, pain, and poignant search."[67]
"The mythic image of Echo," says Kittelson, "suggests a potential for
resonance in a way that 'neurosis' or mindless 'parroting' does not." In
the passionate murmuring of Echo, there is a "calling back, and calling
back, and calling back" that is "full of meaning" and "evokes the kind of
listening that poignantly discovers the significance in Echo's sounds."[68]
    When Kittelson goes on to describe echo experiences as "the singsong
offerings of the psyche," Stevie comes to mind, echoing in his mournful
cadence, "I would *never* do that, *never, never . . .*"[69] For Stevie to stand
present in the text, his words must echo here and now and his oneiric
images gain presence. The word of the child must sound forth beneath
the mute graphic marks that Steven has printed on the page; the dream-
glow of Stevie's image must radiate through the cold abstraction of these
skeletal bodies of light (fig. 5.7).
    It is clear though that, while such a distillation of the text is neces-
sary for the *coniunctio* now at hand, it is not sufficient. More is needed
than Stevie's personal presence. His voiced words and pre-perspectival
images must be Proprioceptively *retracted* in realization of the transpersonal. The Proprioception enacted in the first *coniunctio* is not simply
left behind in thus passing to the second. Rather, the individuation process carried out in the previous chapter remains operative here, since
the presence of the fully conscious Proprioceptive adult is required to
support the Proprioception of the child. Only by sustaining the light of
mature cognitive self-comprehension can a mere regression to mythic
cognition be averted. But exactly what is it that happens here beyond the
personal presencing of Stevie? Just how is Stevie's personal text rendered
transpersonal?
    We know that, for the Proprioception accompanying the first con-
junction (the *unio mentalis*), Steven, having stood present in the text,

begins by proprioceiving his eyes as they view the self-intersection in the stereogram of the Klein bottle (fig. 4.2). In this alchemical meditation, attention then moves further backward from the optical surface of the body into the "blind spot" or hole in the visual/cognitive system. As the process deepens, the hole experienced "in here" links up with the hole in the Kleinian vessel seen "out there," and the Proprioception coalesces into a realization of the (w)holeness of the mental-rational Self, humanity's generic "I" embodied in the neo-mammalian brain. For the Proprioception of *Stevie*, a different meditation on the Kleinian vessel is required—a different Kleinian vessel, in fact.

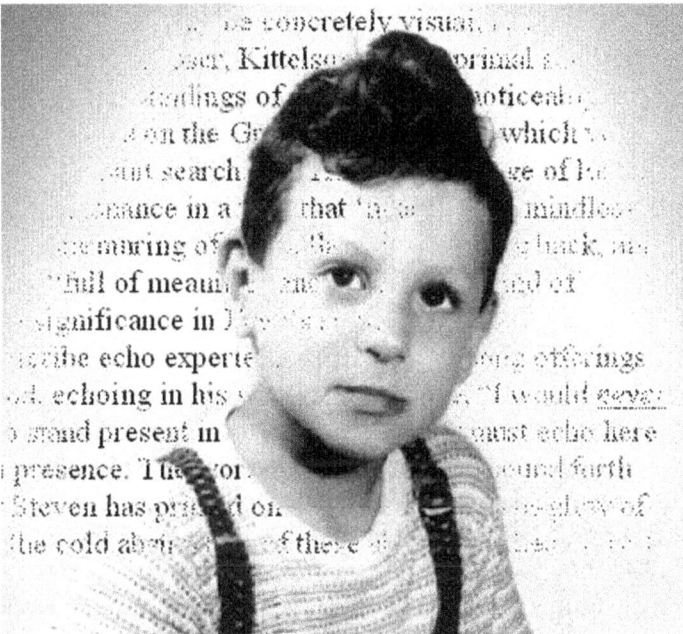

Figure 5.7 Stevie shining through the text.

We have encountered the mythic precursor of the Klein bottle before and are familiar with it: it is the *uroboros*. Should we employ a stereogram of this figure, as we did for its modern, mental-rational counterpart? Stereo-viewing creates a perspectival experience of three-dimensionality appropriate for Steven's post-Renaissance perception, but not for Stevie's pre-perspectival mythic consciousness. Instead of a stereogram then, let us use a classical image of the uroboros (fig. 5.8), with the idea that, when Stevie arrives on the scene, he will not view the figure merely as Steven does—as an object focused upon from a detached perspective point—but will experience it *peripherally*, be oneirically immersed in it.[70]

Figure 5.8 Image of the uroboros adapted from an image appearing in the *Chrysopoeia of Cleopatra*, an ancient Egyptian alchemical text (reprise of fig. 1.1).

With Steven standing behind him to lend support, Stevie must apprehend the uroboros's self-intersection, the place where it swallows itself in an act of paradoxical self-containment. At the same time, Stevie is to direct his attention backward to his eyes. Unlike Steven's eyes, Stevie's are softly focused, moving in the saccadic rhythm of a dream. Guided by the adult Proprioception of Steven that remains transparent for this denser *coagulatio*, Stevie's dream is to become *lucid*; he is to awaken in it to obtain a consciously felt sense of his eyes. Awareness is then to go further backward into the hole in the visual system, moving meditatively from hole to uroboric (w)holeness, from the eyes to the transpersonal "I" of the mythic Self. The movement in this second *coniunctio* follows a Proprioceptive trajectory different from the trajectory of the first *coniunctio*. It is indeed a passage to humanity's communal brain, but a different region of the brain is involved in this second sealing of the Kleinian vessel.

Recall from the last chapter that the distinct anatomical locus of the rational "I" is the "cerebro-ocular" region, that is, the cerebral cortex of the brain and the organ of vision associated with it (the visual cortex is localized in the occipital lobe of the cerebrum). The initial Proprioception entails drawing attention backward into this area of the brain (Burrow's practice of "cotention"). I suggest that, in the denser *coagulatio* now to be enacted, it is not the "I"/eye of the cerebral cortex tied to the written word that is to be Proprioceived, but a *sub*cortical "I" lying behind Stevie's pre-perspectival image. As a matter of fact, a *synesthetic* "I" is operative here, since the backward movement into the Kleinian brain accounts only for the visual or imaginal aspect of the Proprioception, not the vocal. But let us first deal with the former.

The brain research of Burrow was primarily concerned with activity in the cerebral cortex. Almost two decades after Burrow's death, a theory emerged that brought greater emphasis to *lower* centers of the brain. The neurophysiologist Paul D. MacLean proposed that the human brain is *triune*—that it actually consists of *three* brains (fig. 5.9), one arising from the other in the course of phylogeny.[1]

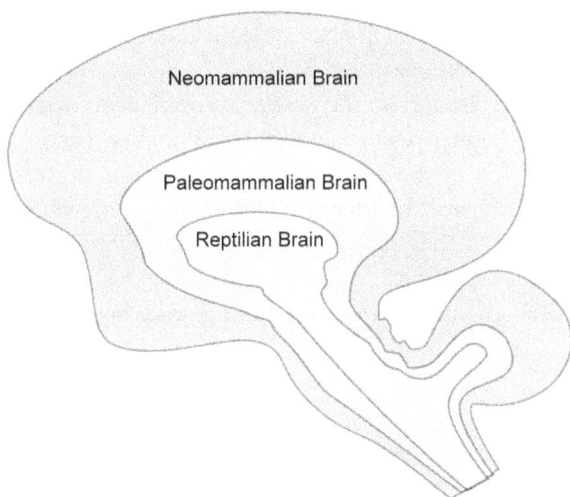

Figure 5.9 Schematic diagram of triune brain. (From Paul D. MacLean, "Alternative Neural Pathways to Violence," in *Alternatives to Violence*, ed. L. Ng [Alexandria, VA: Time-Life, 1968])

MacLean identified the cerebral cortex as the neo-mammalian brain, hypothesizing that it evolved out of an older, paleo-mammalian structure associated with animal functioning. Anatomically, the new brain formed *around* the older brain so that the latter came to be contained within it. The structural containment is also functional. Whereas the neocortex is the layer of the brain in which cognitive activity is consciously processed, the old mammalian brain or limbic system is linked to operations of a subconscious nature. Research using sophisticated imaging techniques (PET studies) confirms that while neocortical activity prevails during waking consciousness, in dreaming the limbic system gains control.[72] The paleo-mammalian brain evolved out of and came to contain an even older structure: the *reptilian brain*. Lying inside the base of the skull, this primal core functions in a largely unconscious, vegetative fashion (the way reptiles appear to function).

I propose that, while the opening Proprioception entails a movement into the neo-mammalian brain, presently it is the paleo-mammalian brain that is to be reentered. The schematic drawing given in figure 5.9 provides us with a rough idea of the spatial relationship among the three brain centers, but a diagram like this is of course an objectification. Here we view the representation of the brain as appearing out in front of us when what is required is a *backward* movement into the brain itself. Entering the paleo-mammalian brain in reverse via Proprioception brings a realization of the brain as the *sub*-objective or psychophysical embodiment of mythic humanity at large.

But Stevie's meditation on the uroboric vessel as his awareness passes backward from dream-eyes to paleo-mammalian brain accounts only for the *imaginal* aspect of the necessary Proprioception, not the vocal. The child's sub-cerebral world does not just involve oneiric visualization but also the voice; it includes sound as well as light. Evidently then, a *dual* Proprioception is called for.

Though Stevie's manner of functioning may be simpler than that of Steven, the structure of Stevie's identity is actually more complex. Whereas the ego of Steven is centered in a unitary core, Stevie possesses an *alter* ego: the presence of "another" is found in his midst. Such a dimorphic feature is to be expected of the young child, since, at this early stage of development, identity has not yet crystallized into the singular form it will later assume (the structure of the child's identity reflects the polymorphous complexity of mythic identity noted at the outset of the previous chapter). Young children often have "imaginary" playmates or companions, a "departure from reality" that adults tend to discourage, or tolerate condescendingly. Quite frequently, the companion takes the form of an animal, whether a creature that is invisible, or an inanimate object that is personified, such as the proverbial teddy bear. The undeniable importance of animals to children recapitulates their great significance in ancient culture, as discussed above. In the dual Proprioception presently required, the Proprioception of Stevie's imaginal "I" is to be coupled with a Proprioception of his liberated animal familiar, his alter-"I." Speaking the language of topological alchemy, I have already intimated that the second stage of the *coniunctio* brings a Kleinian movement backward to the transpersonal core of the mythic human *cogito* that is synchronized with the Moebial backflow of the freed *anima* (the "genie has been let out of the bottle"). I now suggest that, whereas the uroboric Kleinian aspect of the dual Proprioception follows a pathway from eye to limbic brain and culminates in *thinking light*, the trajectory of the Moebial *anima* is from ear to heart and climaxes in *feeling sound*.[73]

The world of primary feeling that "primitive man shares with the animal" is "resonant, layered, slower" than the vibratory world of light; it is "'ear-minded'"; that is, it is a world of sound.[74] And the influence of animals upon the mythic human being evidently occurs through this sonorous channel. Recall Neumann's observation that the gods that presided in the earliest Egyptian dynasties were animals. According to Julian Jaynes (see chapter 1), the Egyptian deities controlled mortal human beings through the power of the *voice*.

In his book, *The Origin of Consciousness in the Breakdown of the Bicameral Mind*, Jaynes hypothesizes that mythic awareness was radically different from the rational consciousness later to emerge. Mythic individuals "do not sit down and think out what they do. They have no conscious minds such as we say we have, and certainly no introspections."[75] Jaynes contends that, for the mythic person, "the gods take the place of consciousness. . . . The beginnings of action are not in conscious plans, reasons, and motives; they are in the actions and speeches of gods."[76] The early human being thus found himself or herself in a largely passive position, at the disposal of external forces personified by the deities: the divinities spoke and mortals listened, acting accordingly. Jaynes goes on to ask:

> Who . . . were these gods that pushed men about like robots and sang epics through their lips? They were voices whose speech and directions could be as distinctly heard . . . as voices are heard by certain epileptic and schizophrenic patients. . . . The gods were organizations of the central nervous system. . . . The gods are what we now call hallucinations. . . .
>
> [In antiquity,] volition, planning, initiative is organized with no consciousness whatever and then "told" to the individual in his familiar language, sometimes with the visual aura of a familiar friend or authority figure or "god," or sometimes as a voice alone.[77]

Later, in a section on the "Authority of Sound," Jaynes raises the "profound question of why such voices are believed, why obeyed."[78] It is because they possess for the listener an overpowering sense of reality. The listener is "somehow face to face with elemental auditory powers, more real than wind or rain or fire, powers that deride and threaten and console, powers that he cannot step back from and see objectively."[79] Viewing the mythic experience of sound as closely akin to the auditory hallucinations of schizophrenics, Jaynes quotes one schizophrenic as reacting to an invisible voice "'as though all parts of me had become ears, with my fingers hearing the words, and my legs, and my head too.'"[80] Toward the end of his book, Jaynes notes that the "phenomenon of imaginary companions in childhood . . . can be regarded as another vestige" of the primal state of affairs.[81] As with the schizophrenic, the unseen companion exerts a highly potent influence on the child through the medium of sound.

To confirm the *animal* origins of the "elemental auditory powers," we turn to the historian of religion Mircea Eliade. In *Myths, Dreams, and Mysteries*, Eliade underscores the significance of animals in early human culture: "animals are charged with a symbolism and a mythology of great importance for the religious life. . . . The prestige of animals in the eyes of the 'primitive' is very considerable; they know the secrets of Life and Nature, they even know the secrets of longevity and immortality."[82] Eliade goes so far as to hypothesize that, in the earliest form of religion, "the Supreme Being may very well [have] manifested himself in the shape of an Animal."[83]

According to Eliade, archaic mythology is often characterized by nostalgia for a bygone "paradisiac epoch" in which people "understood the language of animals and lived at peace with them."[84] If the original intimacy with the animals and their language could be recovered, this would be tantamount to "ascension into Heaven," to meeting "with the gods."[85] We learn from Eliade that the essential role of the *shaman* in mythic societies is precisely to reestablish contact with the "paradisiac" sphere of the animal divinities. Eliade observes that "during his initiation, the shaman is supposed to meet with an animal who reveals to him certain secrets of the craft, or teaches him the *language of the animals*, or who becomes his *familiar spirit*" (note the similarity of this ritual to the totemic initiation rites of the Pacific Northwest described earlier).[86] Once initiated, the shaman can practice his craft by using the "secret language":

> The shaman imitates, on the one hand, the behaviour of the an-
> imals and, on the other, he endeavors to copy their cries. . . .
> The shaman produces mysterious sounds. . . . You think you are
> hearing the plaintive cry of the peewit, mingled with the croak-
> ing of a hawk, interrupted by the whistle of a woodcock: it is the
> shaman who is making these noises by varying the intonation
> of his voice. . . . A good many of the words used in a shamanic
> session have their origin in the imitation of the cries of birds and
> other animals.[87]

So it is sound that is the means by which the shaman taps into the
realm of animal potency, and this suggests that sound was indeed a primary channel—if not *the* primary channel, through which mythic humanity felt the impact of the animal world. That world, I propose, is the
world of "elemental auditory powers." Speaking archetypally, we may
say that the sound wave plays a role for the *anima* similar to that which
the light wave plays for the *cogito*. Sound is thus the archetypal wave-
form of animal action. Furthermore, while the luminosity of the *cogito*
is centered in the brain, the *anima's* sonority finds its rightful place in
the *heart*, for this is where the world of animal feeling has its roots.
Therefore, if the second Proprioceptive conjunction entails a backward
movement from the oneiric eyes that reaches fruition when the mythic *cogito* sees/thinks the light of the limbic brain thereby resealing the
uroboric Kleinian vessel more concretely than in the first *coniunctio*, the
accompanying Proprioception involves a retrograde circulation from the
ears in which the *anima* hears/feels the sound of the heart and thus seals
the Moebius vessel in hermetic fashion. With these archetypal vessels
spinning synchronously backward, human and animal worlds now pul-
sate in an ecological harmony of topo-dimensional Selves.

The question for the now-requisite *coagulatio* is whether this account of dual Proprioception can be distilled more tangibly via the self-
signification of the text. For that, Stevie must certainly stand present.
Given the dual nature of his identity, he will be accompanied by his alter ego, his animal counterpart. In the Proprioception that withdraws
the Projection of Stevie per se, the backward passage of attention from
eye/"I" to limbic brain while contemplating the self-intersection of the
uroboros leads to an encounter with the mythic Self of humanity at large.
In contrast to this visual self-signification, the concomitant Proprioception of Stevie's "double" is essentially auditory: it is concerned with

moving backward through the voiced/heard expression of self-identity into the sonorous animal heart.

While Stevie himself, as a 3 + 1-dimensional being, cannot directly enact the Proprioception in question but must defer to his 2 + 1-dimensional animal counterpart, he may at least be able to pave the way for said Proprioception. I mentioned in the last chapter that work with the Kleinian stereogram involves a kind of mirroring of identity: in viewing the hole in the Klein bottle that leads into the "fourth dimension," I view myself, read my "I." A similar mirroring can be said to occur in the work done with the uroboros, for, in meditating on the uroboros's sub-objective self-penetration, Stevie views his own subjectivity, the imaginal self whose Projection is to be withdrawn. What we presently require in the slower medium of the spoken word is an *auditory* mirroring. Here the mirroring becomes an *echoing* and the body must be proprioceptively entered—not through the visual system, but through the system of speech. Voicing his plaintive "I," Stevie's vocal cords vibrate to produce that word of himself whose sound echoes in his ears. The echo is surely not to be heard in a simply objective way, as if the sound were merely emanating from an external source. What is required instead is what philosopher David Applebaum calls "hearing-immediately."[88] Whereas the "direction of ordinary listening is toward the object, a word-sound . . . hearing-immediately is directed exclusively toward itself, the activity of audition. It thus occupies itself with the inner relations of audition."[89] Emphasizing the processual nature of hearing-immediately (it "pertains to activity, unfolding, doing"), Applebaum relates it to phenomenological philosophy:[90]

> [The] attributes of immediate-hearing . . . remind one of Marcel's analysis of the "body-as-mine." Their commonality allows us to conclude without the shadow of a doubt that it is the management of the lived-body experience, through the means of immediate-hearing, that produces its special ontologizing possibilities. Immediate-hearing converges with the general field of sensation underlying all organic movement, willed or no. The experience of this general field Marcel calls *sentir*; Husserl derives his notion of kinaesthesia from it; others have spoken of it as *proprioception*.[91]

So, with immediate-hearing, we redirect "sensate perceptivity back to the naked bodily whole."[92] In so doing, rather than listening with the "expectation of a result external to our listening . . . we listen to our own listening."[93]

Stevie's auditory mirroring can be related to shamanic practice. Consider Eliade's account of the primary steps in the shaman's journey:

> A shamanic session generally consists of the following items: first, an appeal to the auxiliary spirits, which, more often than not, are those of animals, and a dialogue with them in a *secret language*; secondly, drum-playing and a dance, preparatory to the mystic journey; and thirdly, the trance . . . during which the shaman's soul is believed to have left his body. The objective of every shamanic session is to obtain . . . ecstasy.[94]

To be sure, the Stevie of my childhood did not function as a shaman. Though accompanied by an animal counterpart, he did not intentionally speak the *language* of animals. He voiced the word "I," not an animal sound. By so articulating the "I," the child asserted the human side of his identity that had been developing from birth, and this kept his animal companion on its "leash." To the extent that Stevie was socialized into the world of human signification, his animal double was "domesticated." Of course, in bouts of emotional regression, the child's "I" could quickly dissolve into an animal howl.

But the Stevie of old must not be confused with the one who stands present in the text for the purpose of facilitating Proprioception. With Steven backing him up, Stevie has now come of age and can function not merely with the innocence of the child but with the wisdom reminiscent of the shaman. And the self-signification of his neo-shamanic text unleashes the "elemental auditory powers" of animal vocalization in a non-regressive fashion.

In ritual practices, drumming is often accompanied by repetitive chanting. Our neo-shamanic self-signification operates here as an auditory mirroring that commences with a chanting of the "I." This is not just an echoing of human identity but involves an appeal to the "alter-I." Rather than attempting to "tame" the animal double by imposing the human word upon it as happens in childhood, in the second stage of Proprioceptive conjunction the other is invited to express itself fully in its own language, which is the quintessential language of sound. It is in transforming the spoken human word into a primary animal vocalization that the Proprioception of sound is enabled.

The neo-shamanic practitioner, caught up in the rhythmic echoing of his own voice in synchrony with the drum, can well become ecstatic. Thus Eliade can say that the drum is a "specifically shamanic instrument [that] plays an important part in the preparation for the trance: the shamans both of Siberia and Central Asia say they travel through the air seated upon their drums. We find the same ecstatic technique again among the Bon-po priests of Tibet."[95] Where do these shamans travel in their ecstatic ascents? They pass into the world of primary sound, the realm of echoing animal spirit that they have unleashed. The animals that reside here are hardly docile companions subservient to the human word; they are godlike numinous presences to the shaman of old. Similarly, with the neo-shamanic transformation of the voiced "I" into the basic animal vocalization, the "genie" is unfettered. The sonorous animal into which the chanting author can be transmuted upon striking the animal hide that covers the drum is a "transpersonal, spiritual being. He is transpersonal because, although an animal . . . he is such not as an individual entity, not as a person, but as an idea, a species."[96] In other words, the animal invoked by the neo-shaman's drum-attuned self-chanting is a *topo-dimensional* being; it is the 2 + 1-dimensional Moebial *anima* that signifies itself in the auditory Proprioception of the text-as-sound.

Moreover, the self-signification of sound reverberates in the *heart*. In many a shamanic practice, the fundamental link between the drumbeat and the heartbeat is acknowledged and accentuated.[97] As Lisa Sloan puts it in recounting the experience she had with drumming while writing her doctoral thesis on shamanism, "the drum beat simulated the rhythm of a heart beat. . . . The sound of this beating drum stimulated and resonated with my own heart, suggesting to me that it is indeed through the heart that the shaman sees into the other world."[98]

Let me close this chapter by returning to a theme of great significance for shamanism and alchemy alike: death and resurrection.[99] Eliade notes that shamanic ecstasy can have its share of agony. He describes the initiation of the shaman as involving "a quite complex ecstatic experience, during which the candidate is believed to be tortured, cut to pieces, put to death, and then to return to life. *It is only this initiatory death and resurrection that consecrates a shaman.*"[100] May we suppose that, in the ecstatic rituals described in preceding paragraphs, the shaman is essentially re-enacting the death and resurrection first experienced during initiation? The interpretation seems to be supported by Eliade when he notes that the shaman's ascent, his ecstatic flight into the sky, is equivalent to resurrection: "ascents = resurrection."[101] And this account of shamanism is consonant with portrayals of the *alchemical* opus as involving a series of ordeals that bring the adept to the brink of death and beyond.[102] For present purposes, each Proprioceptive conjunction can be said to entail a shamanic or alchemical ordeal eventuating in the demise of a finite being and its ecstatic rebirth as infinite Self. In fact, we already saw in the last chapter that, in the opening *coniunctio*, Steven faces "decapitation." That is, he faces his demise as a simply autonomous cerebral being detached from others and from nature. Presently, a new alchemical ordeal must be endured in carrying out the second conjunction. Let us call it "death by drowning."

When Steven's abstract rule over this text can be set aside and Stevie can stand present here, the dimension of feeling comes to the fore. Though Stevie—as a human being—is in the first instance a *cognitive* being, the strong influence of the *anima* upon him makes him a creature of passion as well, and the dual Proprioception of the second conjunction carries him and his animal "double" backward into the Kleinian emotional brain and the Moebial heart, respectively. The child cannot survive this process in his familiar form despite the support he receives from his adult counterpart; confronting the awesome *anima*, Stevie will be swept away by a torrent of unbridled feeling. In this regard, an exemplary dream comes to mind.

Dream of April 23, 2010:

After being involved in some kind of risky activity, a gambling venture of some sort, I suddenly encounter an Asian woman. Her elegant, coldly beautiful face seems to radiate intense white light. And she is a cyclops, a creature with an exquisitely mascaraed single eye (fig. 5.10).

The fear I experienced in encountering this stunning *anima* figure was so intense that I woke up screaming. (Was this dream not had by Steven, an adult? The *feeling* in the dream was more the experience of the frightened child within him.) How can such emotion possibly be contained?

A very special vessel is required, as we well know. For his part of the dual Proprioception, Stevie confronts the challenge of sealing the Kleinian vessel in its incarnation as uroboros (fig. 5.8). Moving Proprioceptively backward, the uroboric vessel is realized as the living limbic brain. The turbulent emotional waters are not simply held inside this container but are allowed to pour out completely so that the soul is drawn down

Figure 5.10 Dream cyclops (the image was scanned from a
drawing I made, in conjunction with recording the dream).

into the whirlpool and its erstwhile sense of itself as an independent
being is lost in the undertow. "For souls it is death to become water,"
says Heraclitus, in the cadence of a dream.[D2] Yet, at the same time, the
floodwaters are in fact fully contained within the subtle vessel, hermetically sealed into it through the power of Proprioceptive awareness,
permitting Heraclitus to add, "from water comes soul." Indeed, only a
body of paradox can provide such extraordinary containment.[D3] Recall
Schwartz-Salant's description of the "paradoxical geometry of . . . the
subtle body" in which "'outer' and 'inner' are alternatingly both distinct
and the same." By virtue of the *sameness* of the Kleinian "outer" and
"inner," boundaries dissolve and the individual is swept into the primal
sea where it meets its end. By virtue of the *distinctness* of "outer" and
"inner" (which is upheld by Proprioceptive consciousness), the soul at
once survives its encounter with extinction, though not simply in its
previous form. The resurrected soul both *is* the being it was, and it is
not. No longer merely set off from what lies outside it by well-defined
boundaries, the soul is now itself a body of paradox; a subtle body that
is simultaneously bounded and unbounded, finite and infinite, particular
and universal. The soul is the body of the mythic Kleinian Self.

Therefore, as the dual Proprioception of the second *coniunctio* is carried out, the particular child—supported by the adult's self-signification of the neo-mammalian brain enacted in the first *coniunctio*—moves backward through his uroboric mirror image to the old mammalian brain. In so doing, he suffers the tempest of feeling, succumbs to it, and is reborn as mythic Self. And, in this process, the child's animal companion is unleashed for the co-Proprioception that leads back into the felt depths of the Moebial animal heart (see postscript, below).

At the moment, I, Steven, sit here at my computer experiencing no deluge of feeling—just these lines of text and the cursor blinking back at me. Then let me once again invite the child to stand present on his own for the requisite self-signification. The invitation comes in the form of a dream.

Dream of April 22, 2012:

A man holds a boy in his arms. The man believes he has the child's best interests in mind, but the boy doesn't want to be held, though he doesn't say so. Don't hold the boy against his will, *I inwardly shout,* don't hold him! *But the man continues holding the boy. There's a vague sense that* several people want to free the child. But we can't do that. It's up to the child.

At last the boy openly admits that he doesn't want to be held. So break out!, *I think.* Break out of the confining arms of the man who is holding you. *The boy now starts to become more vocal about not wanting to be held, and I feel relieved to hear him—he doesn't want to be held and he's now* saying *it. And this gives him freedom to break out.* He can break out on his own if he wants, now that he's found his voice.

POSTSCRIPT ON NONHUMAN INDIVIDUATION

I have intimated that the animal undergoes an individuation process of its own, in parallel to human individuation. A full-blown analysis of this is offered in *Topologies of the Flesh*. It was there I first proposed that—since human and nonhuman evolution are interwoven aspects of natural evolution as a whole, a distinct pattern can be identified in which the nonhuman world too has its stages of topo-dimensional individuation. The stages in question are synchronized with those of human development in a close harmony of dimensional spheres reflecting the intimate relationship among the topo-dimensional bodies of paradox (fig. 5.4). Therefore—just as human individuation involves several alchemical conjunctions; several hermetic closures of the Kleinian vessel entailing Proprioceptions of different areas of the human brain and different modes of thinking (rational, mythic, magical, etc.); several ordeals, deaths, and resurrections—so too the animal in its own right undergoes correlated processes entailing conjunctions, closures (of the Moebial vessel), Proprioceptions (of the heart and of feeling), and ordeals. What is more, two other orders of individuation are implicated beyond the human and animal.

However, the developmental concerns of the present work are more focused on the alchemical processes of the human sphere, so I do not repeat the detailed analyses of the stages of nonhuman individuation provided in *Topologies*. To be sure, I do presently deal with the nonhuman realms, an account that includes (1) describing the unique psycho-functional quality that governs each such world (e.g., feeling, in the animal domain), (2) identifying the particular topo-dimensional vessel that contains the individuation of that world, and (3) offering a portrayal of its harmonious interplay with the human world.

DREAM JOURNAL

D1

Dream of November 17, 2010:

I'm out with some friends in a northland region. Everything is frozen and icy, and we're near the sea. There are wild animals around us—sharks or ferocious dogs, something like that. I realize that our lives are at risk. There is the terrifying feeling that at any second we're going to be savagely attacked.

My friends start to run at my urging while I linger behind momentarily. Then, feeling extreme danger from the rear, I start running for my life. A wild animal is following me but we're now in a civilized area and the danger seems to lessen.

D2

Dream of May 24, 2008:

I sit in a chair, looking out at the sea through a large bay window. The ocean vista is beautiful to behold. But now I begin to notice that the sea is rising, and water has begun creeping up toward the base of the house. This alarms me. I dread the prospect of being engulfed.

Presently, the sky clears a little and I am sensing some relief. Though the sea had risen all the way up to the very threshold of the building, it now seems to be receding.

But the relief is short-lived. The sky is darkening again. It is becoming more menacing than before and large waves are forming.

D3

Dream of October 26, 2008:

I view a young man who is crossing a bridge. The bridge does not pass completely over the body of water it spans, but curves down into it.

In watching this scene, I have the strong feeling of viewing a movie that curves back on itself. The film is trying to film itself, I think, and I find that very beautiful. My sense of delight is coupled with the image of a narrow tube bending back on itself like a Klein bottle.

I want my wife to see this too. "Marlene, this is so beautiful. Look at this. Look at this beauty!"

Though the bridge leads the young man down into the water, somehow he is not simply engulfed—as if the bridge were like a Klein bottle, submerging the youth yet keeping the water contained. And my viewing of the scene is also Kleinian in the way it curves back into itself. The beauty and delight I experience from this engages my soul, and I want my *anima* figure to share in it.

CHAPTER SIX

Third Stage of Alchemical Conjunction

Cogito, Anima, and Vegeta

The *coniunctio* now at hand entails bringing the realms of human cognition and animal feeling into Proprioceptive harmony with the denser, more concretely embodied realm of *vegetal sensuality*. In topo-dimensional terms, this conjunction involves synchronizing the 3 + 1-dimensional Kleinian vessel and 2 + 1-dimensional Moebial vessel with the 1 + 1-dimensional vessel corresponding to the next member of our bisection series: the lemniscate (introduced in chapter 3; see fig. 3.5). In the process, the Kleinian vessel is sealed a third time in a still denser medium.[1] I will have more to say about the lemniscate later on. At present, we return to Gebser for a clarification of the inchoate form of thinking that must be Proprioceptively reprised for the third sealing of the Kleinian vessel.

MAGICAL CONSCIOUSNESS

In Gebser's account of the dimensional structures of consciousness, the two-dimensional mythic structure is developmentally preceded by a still more primitive mode of cognition: one-dimensional *magic*. We saw in chapter 2 that magical thinking basically involves thinking by "analogy or association. . . . Magic man feels things which seem to resemble one another as 'sympathetic to,' or 'sympathizing with,' one another."[2] Let us now go more deeply into the underpinnings of magical consciousness.

According to Gebser, the magical structure can be understood in terms of five interrelated characteristics: egolessness, point-like unity with the world, space-timelessness, merging with nature, and magical reaction to the merging.[3] On the first feature, Gebser comments that, with magical consciousness, "the not-yet-centered Ego is dispersed over the world of phenomena" so that everything "still slumbering in the soul is at the outset . . . reflected mirror-like in the outside world." Therefore, "in a sense one may say that in this structure consciousness was not yet *in man himself* [i.e., centered in an individual ego], but still *resting in the world.*" Though he lacked an individual ego, "magic man" did "cope with the earth and the world as a group-ego, sustained by his clan."[4] In illustrating the functioning of the group ego and its magical transactions, Gebser describes the hunting ritual of a tribe of Congo Pygmies in which an antelope is killed not by attacking it directly with bow and arrow but by first drawing a picture of the antelope in the sand:

> With the first ray of sunlight that fell on the sand, they intended to "kill" the antelope. Their first arrow hit the drawing unerringly in the neck. Then they went out to hunt and returned with a slain antelope. Their death-dealing arrow hit the animal in exactly the same spot where, hours before, the other arrow had hit the drawing.[5]

Commenting further, Gebser says:

> In the hunting rite, the *egolessness* is expressed first of all in the fact that the responsibility for the murder, committed by the group-ego against a part of nature, is attributed to a power already felt to be "standing outside": the sun. It is not the Pygmies' arrow that kills, but the first arrow of the sun that falls on the animal, and of which the real arrow is only a symbol. . . . With the Pygmies in their egolessness, the moral consciousness that they must bear responsibility, deriving from a clearly conscious Ego, is still attributed to the sun. Their Ego . . . is still scattered over the world, like the light of the sun.[6]

Gebser views the feature of egolessness as closely related to the second characteristic of magical consciousness: point-like unity. Rather than being centered within a localized, well-bounded ego, magical awareness is "scattered over the world" in a point-like fashion, distributed far and wide in such a way that each point of awareness is immediately linked to every other point. That is why the slaying of the antelope in the sand at one location is inseparable from its slaying in the flesh at another. "This point-related unity in which each and every thing intertwines and is interchangeable, becomes apparent when the symbolic murder in a rite performed before a hunt, coincides exactly with the actual one committed by the hunter."[7] Although the person of magic is "not yet able to recognize the world as a whole but only the details (or 'points')," each of these points "stand[s] for the whole. Hence the magic world is also a world of *pars pro toto*, in which the part can and does stand for the whole. Magic man's reality, his system of associations, are these individual objects, deeds, or events separated from one another like [interchangeable] points in the overall unity."[8]

The third characteristic of the magical structure follows from the second, since, according to Gebser, point-like unity would not be possible in a context governed by space and time. There can be no measurable distance between the Pygmies' sand drawing of the antelope and the forest in which the living antelope dwells, nor can there be a passage of time from the symbolic performance of the ritual killing to the actual. Were it necessary for the nexus between sand and forest to be mediated by spatiotemporal relations, we would not have a direct magical unity but a causal connection between events that are essentially disunited. Therefore, because space and time serve as agents of separation, Gebser concludes that the magical world must be "spaceless" and "timeless."

Although Gebser lists "merging with nature" as the fourth feature of the magical structure, it overlaps considerably with the second feature, point-like unity with the world. Magical consciousness merges and unites with the natural world, is completely woven into it (fig. 6.1). At the very same time, however, magic is used to *counteract* the fusion, and this constitutes the fifth characteristic: "Man replies to the forces streaming toward him with his own corresponding forces: he stands up to Nature. He tries to exorcise her, to guide her; he strives to be independent of her; then he begins to be conscious of his own will."[9] Magical consciousness thus "begins to free itself from its merger with nature, breaking that spell with a counterspell of its own" (as in the antelope hunt described above). This "shattering of ties" bespeaks the "gradual emergence of the ego."[10]

But doesn't Gebser list ego*lessness* as a basic characteristic of the magical structure? Apparently, magical consciousness is not *entirely* egoless, though, in this merged state, the differentiation of person and world, subject and object, self and other, has not yet gone beyond the simplest mirroring. What is implied in the "*pars pro toto*" of magical life is that every part, every other, is a largely undifferentiated reflection of the whole or Self. But parts ("points") and whole do appear here, indicating that a first step forward in fact has taken place, a first objectification, a

Figure 6.1 Magical merging with nature (from a painting
on a chest from the tomb of Tutankhamen, 1300 BCE;
see Gebser, *The Ever-Present Origin*, 53, fig. 5).

first externalization of the relationship between self and other, however
weak. The unitary magical self or subject thus does appear to be bounded, closed within itself and closed off from other, though the boundary
is exceedingly porous. In the next chapter, we will encounter a structure
of consciousness that is more fully egoless, what Gebser calls the *archaic* structure. (Note that Gebser himself, in a table summarizing the
ego's various manifestations across the several structures of consciousness, does not refer to the archaic mode of consciousness as "egoless";
instead, for reasons not entirely clear to me, he prefers to portray it as
"universal" or "cosmic," though he likely would have agreed that it is
also egoless.)[11]

We have just seen that, while Gebser deems the magical structure
to be "egoless," in actuality it possesses an incipient element of ego. A
parallel conclusion can be reached with regard to the other feature of

magical consciousness that Gebser describes in negative terms: "space-timelessness." Though Gebser devotes a chapter to "The Space-Time Constitution of the Structures," he provides little insight into the spatio-temporal constitution of the magical structure as it relates to the more primordial archaic structure.[12] Beyond the bare statement that, whereas the magical structure is "space-timeless," the archaic is "pre-space-timeless," not much clarification is offered as far as I could tell. For my part, I would venture to suggest that the magical world does possess a primordial form of space and time, related, in fact, to its possession of an inchoate form of ego. This revision of Gebser's account would leave the archaic structure alone as essentially spaceless and timeless, just as it is virtually egoless. (In chapter 7, we are actually going to explore a sense in which even the archaic structure possesses ego and spatiotemporality, though not in the positive sense of the other structures.)

In the example of the antelope hunt and elsewhere in his book, Gebser indeed demonstrates that the intimate connectedness of the magical world precludes the linearly measurable space and time that are familiar to us today. Surely then, the magical realm is spaceless and timeless if space and time are taken only in their Cartesian forms. But while space and time entail a displacement from the here and now to an elsewhere and otherwhen, this does not necessarily mean displacement in an extensive linear continuum. All that may be required for such an experience of displacement is the emergence of a fledgling ego to *stand opposed* to nature, establishing nature as what is *other*, something "out there" that is separate from what lies "in here." This opposition between self and other may certainly be weakly developed; the magical human being may be intimately united with the other, as in the hunter-gatherer experience of the wild animal described by Berman (see previous chapter). Yet it appears this would be enough to open up a sense of displacement critical to an experience of space and time. I am proposing then, that the experience of the "intimate other" prevailing in the magical world brings a nascent experience of space and time.

Let me end this section by considering space and time from a somewhat broader perspective. Bearing in mind our association in chapter 5 of time with subjectivity and space with the objective, it seems we can generally say that the more subject and object (self and other) are opposed to or differentiated from one another, the greater should be the differentiation of time and space. In the transition from magical to mythic consciousness, "man" extricates himself "from his intertwining with nature" and there is an "emergent awareness of the internal world of the soul," this in opposition to the world outside.[13] So mythic consciousness brings a heightened sense of subject-object, inner-outer differentiation, and, concomitantly, a sharpening of the distinction between time and space. Of course, the *sharpest* distinction between time and space must await the subsequent emergence of the mental-rational structure of consciousness. The observations of eco-philosopher David Abram are relevant to all this.

In *The Spell of the Sensuous*, Abram makes it clear that indigenous peoples did not perceive the categorical separation of space and time known to us today. Drawing from anthropological studies of the Hopi, Navaho, and Aztecs, Abram emphasizes the interwovenness of space and time for these older cultures:

> Unlike linear time, time conceived as cyclical [as Gebser said of mythic time] cannot be readily abstracted from the spatial phenomena that exemplify it—from, for instance, the circular trajectories of the sun, the moon, and the stars. . . . Thus cyclical time, the experiential time of an oral culture, has the same shape as perceivable space. And the two circles are, in truth, one.[14]

This is not to say that Abram denies any distinction between space and time in mythic culture:

> While [anthropologist Benjamin] Whorf did not find separable notions of space and time among the Hopi, he did discern, in the Hopi language, a distinction between two basic modalities of existence, which he terms the "manifested" and the "manifesting." The "manifested" corresponds roughly to our notion of "objective" existence. . . . [to] that aspect of phenomena already evident to our senses. . . . The "manifesting," on the other hand, "comprises all that we call future . . . [as well as] everything that appears or exists in the mind, or, as the Hopi would prefer to say, in the *heart*."[15]

We may surmise that the "manifested" dimension—given its correspondence to "'objective' existence," is the forerunner of what we presently take as space, whereas the "manifesting," being concerned with the future and with the mind or heart, is a precursor of modern temporality. But, again, the "manifested" and "manifesting" are closely entwined in mythic consciousness. Only with the coming to dominance of abstract mental-rational consciousness do space and time appear sharply divided from each other. Space now stretches out before us as an extensive three-dimensional continuum in which we have freedom of movement in all directions. Time, on the other hand, is an *intensive* dimension; we are not at liberty to move around in time but can only experience its unidirectional flow from the past into the future. In the most general terms then, we can conclude that the differentiation of space and time is progressively enhanced with the movement to more developed structures of consciousness.

VEGETAL SENSUALITY

Magical consciousness is the mode of experience that must be Proprioceptively reprised for the third *coniunctio,* the third sealing of the 3 + 1-dimensional Kleinian vessel. Having now elucidated this primordial form of human awareness, it is time for us to consider its associated nonhuman milieu from the topo-alchemical standpoint that will bring into play the lower dimension of vegetal sensuality. Remember as we proceed that while Gebser took magical consciousness itself to be one-dimensional, here we are regarding it as 3 + 1-dimensional since it does constitute an incipient variety of human cognition. We are going to see that it is magical cognition's nonhuman counterpart that is actually one-dimensional, or rather, 1 + 1-dimensional, taking into account its weakly differentiated temporal dimension. (The space-time notation I am using, "1 + 1," of course does not do justice to the near inseparability of space and time in the lower-dimensional sphere.)

In the previous chapter, I alluded to the *anima mundi* or "world soul," a concept "founded on the view that the [nonhuman] world is productive of life and animation, and can therefore be regarded as itself animate."[16] This general notion of nonhuman vitality is not limited to the animal sphere but encompasses the entirety of nature. An etymological fact is relevant here. The word *animal* is of course linked with the idea of "animation," but so is the word *vegetable.* In its adjectival form, *animal* derives from the Latin *animalis,* "living," "animate," and "vegetable" is "from LL. *vegetabilis,* animating, hence, full of life, from L. *vegetare,* to enliven, quicken" (*Webster's [Unabridged] Dictionary*). Bearing in mind Jung's intimation that the freeing of the *anima mundi* from cognitive human control is a primary goal of alchemy, we turn our attention to the vegetable aspect of the *anima mundi.* From this chapter and the last, we know that both mythic and magical societies were closely attuned to the emotional realm of the animal. What may we say of their relationship to the plant world?

In Berman's commentary on prehistoric society, he has much to say about human ties to the animal kingdom, but says very little about the relationship to plants. This is not surprising given his primary concern with human affairs. Human beings generally do have a more immediate connection with animals than with plants inasmuch as they are closer to animals on the phylogenetic scale. Yet the great significance of vegetable life in the prehistoric world can hardly be denied. Eliade explores the spiritual aspect of this in his study of shamanism.

As we found in chapter 5, the shaman's goal is an ecstatic return to the bygone "paradisiac epoch."[17] While the journey is often guided by animal spirits or deities, animals are not the only beings who serve in this capacity. According to Eliade, "In South America, as everywhere else, helping spirits can be of various kinds: souls of ancestral shamans, spirits of plants or animals."[18] In fact, plants commonly play an important role in many aspects of shamanic practice. Rituals quite frequently involve the burning of plants like juniper and thyme as incense.[19] The pungent odor of the smoke can have an intoxicating effect; it can help generate the "magical heat" that facilitates the shaman's entry into the state of ecstasy.[20] Eliade notes that many practitioners "eat highly aromatic plants; they hope thus to increase their inner 'heat.'"[21] Related to this is the "ecstatic function of the [hot] vapor bath, combined with intoxication from hemp smoke, among the Scythians."[22] "There is every reason to believe," says Eliade, "that the use of narcotics was encouraged by the quest for 'magical heat.' The smoke from certain herbs, the 'combustion' of certain plants had the virtue of increasing 'power.' The narcotized person 'grows hot'; narcotic intoxication is 'burning.' . . . Often the shamanic ecstasy is not attained until after the shaman is 'heated.'"[23]

Typically, the shaman's ecstatic journey reaches its climax with "magical flights" to "Heaven" or the "celestial spheres." The critical importance of the *vegeta* in these trips aloft is symbolically expressed by the fact that they are frequently enacted by climbing "the Cosmic Tree [that] is supposed to be situated at the Centre of the World and that connects Earth with Heaven."[24] Eliade describes one such ritual at length.[25] The Altaic shaman sets up a young birch tree in the middle of the yurt (ceremonial hut), positioning it so that it protrudes through the smoke hole in the top. Notches are cut in the tree as footholds for the ascent to Heaven. The ritual fire is lit, and, after sacrificing a horse, the shaman begins his ecstatic climb. As he goes up, he becomes more and more heated. With his agitation mounting, he beats his drum violently, loudly intoning nonhuman sounds to announce the presence of accompanying animal spirits. Whereas the skin stretched over the drum comes from the hide of an animal, its barrel is fashioned from one of the branches of the Cosmic Tree. As Eliade explains in his more abbreviated account of the Altaic ceremony, since "the frame of his drum [is] made of the actual wood of the Cosmic Tree, the shaman's drumming transports him magically near to the Tree; that is, the Centre of the World, the place where there is a possibility of passing from one cosmic level to another."[26] Contributing to this "vegetable magic" is the periodic burning of incense (juniper). With the ceremony reaching its climax, the shaman encounters "Bai Ulgän, the supreme God," and exits rapturously via the same hole atop the yurt through which the intoxicating smoke drifts out.[27]

Despite the central importance of plants in the practices of shamans, it could still be argued that the prehistoric relationship of human beings to the vegetable realm is surpassed by the stronger relationship to the animal. This might be taken as especially true for the magical consciousness of the Paleolithic hunter-gatherer. We can infer from Berman's description of Stone Age society that, while the Neolithic link to the animal sphere was certainly close, it was not quite as intimate as its Paleolithic predecessor. Again, in the Paleolithic Age "animal life was everywhere, even in the skies," and "animal movement, the animal body, was the model of human expression."[28] In hunter-gatherer societies, says Berman, "human life . . . had no special significance apart from the animal world."[29] However, when we turn to Gebser's observations on the experience of the hunter-gatherers, we learn that their magical unity with animals had a decidedly *plant-like* quality!

Gebser characterizes the magical realm as basically *vegetative* in nature: "the realm of vegetative energy," that of the "vegetative intertwining of all living things."[30] In the same vein, Gebser speaks of "the late drawings, paintings, and frescoes of magic man, in which man's merger with nature is portrayed so graphically that the entire picture is nothing but a plant-like amalgam" (see fig. 6.1).[31] Elsewhere Gebser alludes to magic man's "merger with the primeval forest."[32] The conclusion reached in the previous chapter stands: mythic consciousness is indeed closely linked to the animal domain. But what we are now seeing is that magical consciousness is at bottom more plant-like than animal. Whereas mythic awareness is well synchronized with the rhythm and flow of the animal world, the point-like world-entanglement of magical cognition that Gebser describes is essentially expressive of the vegetative domain. Does this distinction have psycho-functional implications?

Jung's developmental perspective on the four basic functions of the psyche was set forth in chapter 4 and then related to the sequence of alchemical conjunctions wherein the psychic functions are reprised in reverse order of their emergence. The first conjunction involves the *unio mentalis* and centers on the primarily human function of thinking. Next is the *coniunctio* that draws us Proprioceptively backward into the older, more concrete realm of mythic feeling, which we have related to the world of the animal. Should the third conjunction not bring us back into the still more primal domain of the senses? Should it not be sensation and sensuality more than feeling that prevail for magical consciousness and the plant world?

The magical world is indeed the realm of the senses, but this does not refer to the sublimated, constricted form of sensation controlled by the wakeful *cogito*. For magical experience is such that reflective consciousness is *asleep*, says Gebser.[33] Yet he can still speak of the "sharper senses" that prevail in the "one-dimensional" action sphere, of the "heightened, natural, sensory apperception of magic man—superior to our own."[34] Evidently, the form of sensibility holding sway in the one-dimensional milieu entails the vital vegetative force of life, of sensuality and sexuality. The sense experience of the one-dimensional domain is driven by impulse and instinct, rather than by a higher reflectiveness, and these biological directives, in turn, are related to sexuality.[35] Time and again, Gebser associates the one-dimensional world with *vitality*, life-energy or life-force. Whereas water is the operative element for the two-dimensional mythic soul, "blood and semen . . . [are the] pre-eminent vital forces" of magical one-dimensionality.[36]

For Gebser, an important correlate of a structure of consciousness is its "organ emphasis." In the last chapter, we saw him associate mythic consciousness with the heart. Gebser would relate the magic structure to the viscera or intestines: "viscera represent the intertwined unity which we have ascertained as characteristic of the magic structure of consciousness."[37] No organ is specified for the archaic sphere. Conspicuous by their absence from Gebser's account are the *genitalia*. Given the strong link Gebser establishes between magic and sexuality, one would think that the region of the body accentuated in magical functioning might be the groin more than the intestines or guts. And, indeed, there is an element of ambiguity in Gebser that opens the possibility of relating the magic structure to the genital organs rather than the intestines. In discussing the latter, Gebser says that they "are evidence of unity *or identity.*"[38] But elsewhere in his text, Gebser brings out the fundamental *distinction* between unity and identity, defining "magic man" in terms of the former: "he is distinguishable above all by his transition from a zero-dimensional [archaic] structure of *identity* to one-dimensional *unity.*"[39] Then could it not be that the intestines would be the organs emphasized in the archaic identity structure, leaving room for us to view the genital organs as those of unitary magic? I might be making too much of the phrase, "unity *or identity.*" But it does seem that, for Gebser, an obvious relationship should exist between the magic structure and the genitals, that the sexual organs are nonetheless conspicuously absent from his account, and that he fails to provide an organ correlate for the archaic structure. Perhaps then we should amend Gebser's analysis accordingly, linking the magical domain to the reproductive organs and the archaic realm to the intestines.

But a further word of qualification is in order here. In correlating structures of human consciousness with bodily organ centers, Gebser did not sufficiently take into account the distinction between human and nonhuman worlds. Consider the relationship between mythic thinking and the heart. We know that, *as* a form of thinking, the mythic structure has a primary relationship to an organ center located well above the heart, namely, the old mammalian brain (or limbic system). In chapter 5 we found that—while the mythic human being is certainly strongly influenced by the soundings of the heart—it is not the human *cogito* as such who stands in primary relationship to the heart but the non-human *anima*. So rather than simply associating mythic thinking with the heart as Gebser did, it seems we should associate it primarily with the paleo-mammalian brain, while acknowledging the potent influence of the animal heart on mythic humanity via the human heart. We arrive at a similar conclusion for magical thinking: its *primary* organ correlate is the *reptilian brain* (to be further discussed below), though the magical human being is powerfully affected by the sub-cranial reproductive center, whose own primary relationship is with the *vegeta*. In the next chapter, we will explore the relationship of *archaic* thinking to a fundamental brain structure called the *neural chassis*, to the intestines, and to our ultimate topo-dimensional structure, what we shall call the *minera*.

Before seeking to confirm the lower dimensionality of the vegetative sphere, let me reiterate my departure from Gebser on a related matter also stemming from his tendency to overlook the nonhuman world as such: the question of dimension number. The proposition I am offering is that we consider magical cognition per se not as one-dimensional, but as an inchoate form of 3 + 1-dimensional human consciousness embedded in a 1 + 1-dimensional nonhuman environment of simple vegetative life. The latter realm corresponds to a more primal aspect of the *anima mundi* than the animal's emotional world; it is a world associated with reproduction and sensuous life-energy. The suggestion then is that, if the 3 + 1-dimensional mythic human being was intimately attuned to the 2 + 1-dimensional realm of animal emotion, at the earlier, magical stage in the development of the 3 + 1-dimensional *cogito*, there was an attunement to the 1 + 1-dimensional "vegetative-vital" order of nature, the sphere of primordial sensuality, impulse, and instinct.

Now, in the previous chapter I cited the work of Lachelier as indicating that animal consciousness is essentially emotional and is played out in two-dimensional space. While he was not as explicit about vegetable consciousness, the implication is clear that, indeed, it would be one-dimensional. For Lachelier set forth a dimensional hierarchy of being, from line to plane to three-dimensional space; he described the hierarchy of nature, from mineral to vegetable to animal to human; and he explicitly associated humans with three-dimensionality and animals with two-dimensionality.[40] It therefore follows that vegetables must be one-dimensional (what we consider to be 1 + 1-dimensional). But did Lachelier relate the vegetative sphere primarily to sensation and sensuality, as we have done? He did not.

First of all, while Lachelier identifies the two-dimensional animal world as one suffused with emotion, he appears to combine or conflate emotion with sensation. He says, for example, "the *self* is at once the will to live and the fundamental emotional state. . . . Such is . . . our sensuous *self* or the animal *self* in us."[41] Similarly, Lachelier associates sensation with "desire, or purpose," and the latter, in turn, with "two-dimensional space or surface."[42] As for the one-dimensional sphere, Lachelier claims that "the vegetable has no outer senses and nothing external can exist for it. There is, therefore, room in its consciousness only for the obscure feelings which doubtless express in it the slow evolution of nutritive and reproductive tendencies."[43] So Lachelier evidently would conclude that while the feelings and sensations of the two-dimensional animal world are well developed, in the one-dimensional vegetative realm, they are faint and incipient.

In the foregoing account, Lachelier seems to presuppose a dualism of inner and outer reality. Because the "vegetable has no *outer* senses and nothing *external* can exist for it," the assumption appears to be that it is

limited to an insular, *inner* world where there is "room in its consciousness only for . . . obscure feelings." Apparently then, concrete experiences depend entirely on exposure to outer reality: without external senses,
there can be little or no sensation or feeling. In contrast to this view, what
I am proposing is that the very division between inner and outer reality
only first arises in earnest with the emergence of 2 + 1-dimensional being, and that, in this domain, sensation is subordinated to feeling. To be
sure, we can say with Lachelier that the "vegetable has no outer senses,"
but this hardly means that it simply does not sense. Rather, vegetative
consciousness would be *less* limited than animal consciousness in this
regard. The world of the *vegeta* is governed by the relation of *pars pro
toto* that Gebser ascribed to magical consciousness, the relation of total
intertwinement that precedes the articulation of the inner-outer split. It
is this split that calls forth emotion, for emotion requires a psyche, an
individualized soul over against which there is cast an exterior world;
when the division occurs, mature emotion takes precedence over sensation, just as mature cognition rules both sensation and emotion in the 3
+ 1-dimensional domain. But *prior* to the division, there is *sheer sensuality*, an all-encompassing vital participation in a world that—far from
being external to consciousness, is fused with it.

The completely sensuous, vegetative character of the 1 + 1-
dimensional sphere is intimated by Ouspensky. In the last chapter, we
heard his conclusion that the animal realm is governed by emotion and
is two-dimensional. With respect to the lower-dimensional domain,
he says: "For the being possessing sensations only, the world is one-
dimensional."[44] Although Ouspensky does not go into detail on the *vegetative* nature of the one-dimensional sphere, he does note the following: "Those influences which proved to be beneficial for a given species
during the vegetative life, with the transition to the more active and
complex animal life begin to be sensed as *pleasant*, the detrimental influences as unpleasant."[45] The implication is that while the two-dimensional
sphere of animality is ruled by emotion (the experience of pleasure and
displeasure being fundamental), the lower-dimensional realm of the vegetable is pre-emotional; indeed, it is purely sensuous.

It should be clear by now that merely writing *about* vegetal dimensionality and magical consciousness will not by itself bring about the actual
coniunctio with which this chapter is concerned. What is needed in order
to seal the Kleinian vessel a third time and liberate the *vegeta* is a new
and still more concrete act of *self-signification*. Before proceeding with

this, we turn once more to the question of topological containment. If the cognitive Self is a 3 + 1-dimensional being whose individuation is contained by the Klein bottle, and if the emotive Self is a 2 + 1-dimensional being whose containment vessel is the Moebius, how may we describe the topological vessel that contains the 1 + 1-dimensional sensuous Self of the *vegeta*? The answer is found in the next member of our bisection series: the *lemniscate* (fig. 6.2).

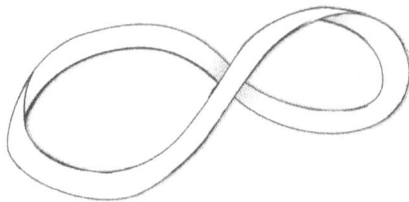

Figure 6.2 Lemniscate.

Just as bisecting the Klein bottle yields the Moebius structure, cutting a Moebius down the middle will produce the lemniscate. We studied this closely knit topological family in chapter 5 (see fig. 5.4) and established that—despite its appearance to conventional mathematics as but a series of two-dimensional objects cast in higher-dimensional space—in fact the series constitutes a grouping of spatiotemporal, sub-objective dimensions unto themselves. Though these worlds are strongly interrelated, each embodies a distinct order of psychophysical Selfhood and each develops dynamically, undergoing its own process of individuation. We can now confirm that, whereas the 3 + 1-dimensional Kleinian order entails human cognition and the 2 + 1-dimensional Moebial order embodies animal emotion, the 1 + 1-dimensional lemniscatory Self is that of vegetal sensuality.

In the course of human history, the transition from magical to mythic culture obscured the earlier intimacy of humanity with the vegetable realm. The topo-dimensional correlate of this repression of "vegetable magic" is the closing of the 1 + 1-dimensional lemniscate into the individuating 3 + 1-dimensional Kleinian vessel. This "bottling up" of the exuberant sensuality of the *vegeta* by the developing *cogito* was then followed by its even deeper repression, as the Kleinian Self became further individuated and human culture advanced into the Greek and post-Renaissance epochs ruled by abstract human rationality. For the lemniscatory *vegeta* to be liberated, the Kleinian vessel must be sealed hermetically a third time. To the *unio mentalis* and *unio emotionalis* must be added a *unio sensualis,* as it were. Here Proprioceptive awareness reaches all the way back into the inchoate mist of magical mentation that predominated in the epoch governed by the sensuous *vegeta.* And resealing the 3 + 1-dimensional vessel in this fashion allows the lemniscatory *vegeta* to free itself from repressive containment and enter into topo-dimensional harmony with its Kleinian counterpart.

Turning to the *vegeta*'s relationship with the Moebial *anima,* the latter undergoes its own process of individuation in the course of which the lemniscate comes to be repressively contained within *it* (the lemniscate is thus doubly repressed). The Moebial vessel therefore must also be resealed for the complete liberation of the lemniscatory *vegeta.* This will entail a second Proprioception on the part of the *anima,* carrying it into a deeper, more primordial recess of the heart. Said Proprioception will free the *vegeta* from animal containment and bring *anima* and *vegeta* into harmony. Topo-dimensional harmony will thus be realized by respective pairings of the liberated lemniscatory vessel with Kleinian and Moebial vessels. Adding to this the harmony of the Moebial and Kleinian pair brought about in the second Kleinian *coniunctio* (described in chapter 5), all three topo-dimensional vessels (subtle bodies, Selves) will now be spinning Proprioceptively backward in synchronized pairs (I will touch on the *vegeta*'s own Proprioception in the next section). The question remains, of course, as to what we must do to *concretely enact* the new Proprioceptions right here in this text. It seems additional self-significations are called for.

SELF-SIGNIFICATION OF THE MAGICAL INFANT

My intention is to turn these abstract words about alchemical harmonies and conjunctions into a living reality by bringing to attention not just the signified content of the words but the concrete *process* of signification as well. In previous chapters, such self-signification has been facilitated by fleshing out the text via Kleinian signifiers. The aim was also to include in the text the embodied self of him who signifies: the author of this work. In my attempt at mental self-signification (the *unio mentalis* of chapter 4), it was essential for me to drop my cloak of anonymity and make my presence tangibly felt. I, Steven Rosen, was to stand present on the page, and having unveiled myself, it was then necessary for the Self's transpersonal Projection of "Steven" to be withdrawn by a Proprioceptive movement backward into the neo-mammalian brain of humanity at large, a process that could be aided by alchemical meditation on the stereoscopic Kleinian signifier (fig. 4.2). To the extent that the Proprioception has actually taken place, it has allowed the generic Kleinian Self to assume authorship of the text.

Next came *mythic* self-signification (chapter 5) and the text was to be further coagulated by calling to presence the long-repressed, subtextual aspect of author Steven's identity: *Stevie*, the oneiric child. The subsequent Proprioception of this youth, assisted by imaginal meditation on the uroboric precursor of the Klein bottle (fig. 5.8), sought to follow a backward trajectory into humanity's paleo-mammalian brain, thereby further consolidating Kleinian authorship.

What we presently require is an even more concrete self-signification involving the *magical* consciousness of a still more deeply repressed facet of identity. This primal child dwells in a dense realm of vegetal energy; a domain of blood, semen, and fire; of wild instinct and raw sensuality. I call the subtextual being who must now stand present "Baby Stevie." The following dreams are intended to encourage the infant's appearance. No doubt they contain much that is of a personal nature, much that could serve as grist for the interpretive mill. But it is the *prepersonal* aspects of the dreams that are of greater interest in the present context: the images of blood and fire, sexuality and mayhem, that evoke the darkness of the *vegeta*.

Dream of January 4, 2003:

> In a secretive way, a woman has brought about carnage and the blood from her victims has crystallized into a red dress snugly sheathing her body. In the closing image she lies on her side in a voluptuous pose that brings to my mind Goya's famous painting, The Nude Maja. This image then transmutes into the one shown in the trailer of the film American Beauty depicting a naked young temptress covered with red rose petals.

Dream of December 10, 1987, "The Woman Inside/The Woman Outside":

> A woman is imprisoned behind a brick wall. She is being held for the performance of a dreadful ritual: she is to be burned alive. I am in a state of terrible anguish about the unspeakable agony the victim is to undergo. More than anguish. The thought of someone going through this hellish process, the thought of what she is to experience, is something just too palpably awful for me to bear.

An older woman stands outside the brick wall and prepares to give the command to begin the ceremony. The woman outside is adamant, solemnly insistent that the ritual be carried out. Her heart is hardened to the ordeal the woman inside is about to endure. I find this intolerable.

What I feel in this dream is a sense of astonishment, complete disbelief that any person could undergo such suffering, and that another could have the heartlessness to inflict it. And I am overwhelmed by an uncanny sense of the impossible: how can an event so totally unthinkable actually be coming to pass? But the dream also has an undertone of solemn dignity that accompanies the incomprehensible cruelty. A primal ritual is being enacted that is beyond my capacity to accept or understand.

Dream of December 22, 2009:

I sense myself embedded in some kind of dense matrix of people and the feeling is at once numinous and ominous.

At some point, I leave the group and go off to the right, into a garden suffused with a yellowish aura. Here I encounter a large insect that has the circular shape of a flower, a dandelion maybe. Or is it the insect itself that is yellow? The sight of this creature lighting on a bush repels me. There is a blurring together of the insect with the petals of the flower that strikes me as somehow menacing and obscene.

Dream of July 5, 2008:

This is a dream of naked horror. Order has broken down and people are killing each other. Hordes of people are just murdering each other in the streets, seemingly at random. It's hell.

I'm about to be killed. Someone has raised a cane or a pipe and is on the verge of demolishing me. But then, this little Hispanic kid, a child named Juan, saves me, even though he too is quite wild.

I'm very grateful to Juan. I want to go to his home to thank him. But he lives in a dangerous Hispanic neighborhood where the residents might attack me just because I'm white. I suggest to Juan that he meet me halfway but he blows it off. He's raw, feral, wild. A real wild boy.

Still, I have the impression he'd like me to travel to his place. Being very grateful to him—and also somewhat fearful of this little wild boy—I say I'll come there.

I'll give him a lot of money. Will that do it? What can I give him to thank him for saving my life—and to stay on his good side?

The general feeling of the dream is one of danger, danger, danger: extreme danger at every turn. People are killing each other indiscriminately and there's an imminent threat it will happen to me.

And yet, in the midst of all this, I find myself in an erotic encounter at the end of the dream. A woman and I are viscerally drawn to each other. The chemistry is strong and we want to make love. Where can we do it amidst the pandemonium?

In the last dream of the set, Steven's infantile counterpart turns up in the guise of "Juan," the feral child. This is the wild boy within me, the primal infant, "Baby Stevie." I said in the previous chapter that once Stevie stands present and pours his heart into the text via the medium of dreams, his Projection as a finite particular being must then be withdrawn. Through Proprioception, the emotion-filled cognitions of this particular child are transformed into the *generic* cognition of mythic humanity at large. In the case of *Baby* Stevie, there is no well-established sense of personal identity that needs to be withdrawn and rendered generic through Proprioception. While the human infant is certainly regarded as an individual being when seen through the differentiating eyes of the adult, the baby itself is largely entangled in the communal identity of the magical world, and its individuality is but weakly Projected. Therefore, when "Baby Stevie" comes to presence, no significant Proprioceptive effort is required that would return the child to a communal level of cognition. What *is* required before being able to carry out the minimal Proprioception in this *coniunctio* is a distillation of Steven's written text into a subtextual medium that is even denser than the concrete images and spoken words of Stevie. I am referring to the medium Jung associated with the psychological function most deeply immersed in the unconscious: *intuition.*

The magical world is darker than the mythic, entailing a deeper state of sleep.[46] The subject of this world does not have an abstract concept of himself or an image of himself. The prime signifier of his text is neither the written "I" nor the visual self-image but only a basal intuition of his own being. Since this dim apprehension is all that supports the elemental operations of consciousness here, it must suffice as the basis for the text. So when Baby Stevie stands present to engage in self-signification, the "master signifier" of his magical text will be a self-*intuition*, not a self-image. And this rudimentary intuition of his own existence is what Baby Stevie must Proprioceive (with Steven's assistance, of course).

If Steven's Proprioception involves the passage backward through the abstract "I" to the innermost core of the neo-mammalian brain; and if Stevie's Proprioception entails the retrograde movement through his self-image to the paleo-mammalian brain (limbic system); then Baby Stevie's self-realization should require moving in reverse through the intuition of himself to the generic *reptilian* brain. In the preceding chapter we found that the so-called reptilian brain is the most primitive member of MacLean's triune grouping. This deeply embedded cranial structure operates in the largely unconscious, vegetative fashion characteristic of Gebser's magical structure. Is there a contradiction in saying that what is "unconscious" and "vegetative" nonetheless possesses vitality? We already know there is not. According to Gebser, even though *rational* consciousness is inactive in the magical world, that world has its own kind of dynamic; again, it is the "realm of vegetative energy" in which "each and every thing intertwines and is interchangeable."⁴⁷ Remembering also the etymology of the word *vegetable*—"from LL. *vegetabilis*, animating, hence, full of life" (*Webster's [Unabridged] Dictionary*)—it seems clear that the magical reptilian brain to be Proprioceived is by no means simply lifeless and inert. (Although magical thinking is indeed "vegetative" under the influence of the *vegeta*, let's not forget the *distinction* between the magical realm and that of the *vegeta* per se: the former is a 3 + 1-dimensional sphere of incipient human cognition whose primary locus of operations is found in the brain, whereas the latter is a 1 + 1-dimensional domain of nonhuman sensuality chiefly associated with organs of reproduction.)

While images of the Klein bottle and its uroboric precursor have been employed meditatively to directly facilitate the processes of mental and mythic Proprioception (respectively), an image cannot similarly be used for magical Proprioception, since said Proprioception is not visual: it does not involve a movement from the optical surface of the body to the brain. We know, in fact, that magical consciousness is generally undifferentiated. Consistent with its lack of differentiation between subject and object (signifier and signified), self and other, and inside and outside, is a lack of differentiation between the outer surface of the body and its interior recesses—the boundaries of *anatomical* space are also undeveloped in the weakly spatiotemporal world of magic. This suggests that in Proprioceiving the infant's intuitive text, the apprehension of the lived reptilian brain—rather than having to be mediated by a pathway extending from a well-defined surface of the body to its internal core—is realized through immediate intuition. With this, the Kleinian vessel is hermetically sealed for the third time, though a visual image of the

Klein bottle does not appear in the lightless magical realm. Of course, for magical Proprioception to occur, the infant must be present in the text, and—with the support of Steven—he must awaken from his deep sleep to render the process conscious. Without the infant's tangible presence here, we can only speak abstractly.

When Baby Stevie does arrive on the scene, he will not be alone. We saw in the last chapter that Stevie is accompanied by an animal companion. In the case of the infant, *two* nonhuman companions attend. It is not surprising that Baby Stevie should possess one more alter ego than does Stevie, given the fact that the infant's identity is less cohesive than the older child's. While the latter can express to us in words and deeds the nature of his "imaginary friend," the baby is not so articulate. With respect to the magical infant's first nonhuman counterpart, I suggest that, like the mythic child's companion, it is an animal, but one that is more unremittingly feral than the older child's alter ego. This is consistent with the conclusions reached about magical and mythic culture in the previous chapter. In the Paleolithic Age, the magic-minded hunter-gatherers lived in ongoing intimacy with wild animals, creatures that could pose a threat to life and limb at any moment, creatures with whom the Paleolithic person was deeply identified. In making the transition to the New Stone Age, the establishment of agricultural communities lessened the immediate physical danger from wild animals. And the Neolithic domestication of animals (which brought about the wild-tame distinction discussed in chapter 5) must also have mitigated the overpowering effects of the primitive animal on the human psyche. This is not to say that the wild animal was no longer a formidable figure for human beings—not as long as such animals could still be worshipped as gods.[48] The ambiguous and ambivalent nature of mythic culture's relationship to animals is ontogenetically recapitulated in developmental psychologist Selma Fraiberg's classical account of "Laughing Tiger": A little girl copes imaginatively with her intense fear of wild animals by inventing as her playmate a docile tiger that cannot roar, but can only laugh (I don't recall Stevie having a companion of this sort, but it is possible he did).[49] Far less ambiguity exists in the *magical* relationship to animals. The untamable alter ego of the magical baby is more persistently fearsome and ferocious—the "intimate alien" that Berman alluded to.

Of course, from the topo-alchemical standpoint, the transformation of humanity's relationship to the animal world could not simply have resulted from the development of domestication practices by human beings. I have indicated that the Kleinian individuation of the human being is intertwined with the Moebial *anima's* own individuation process. Is there any evidence in the archaeological record that might be consistent with this? Anthropological research suggests that animals were not merely domesticated by human beings, but were capable of domesticating themselves.[50] An example of this is the evolution of dogs into loving and faithful companions from what were wild wolves. Biologists Raymond and Lorna Coppinger hypothesize that this transformation was brought about by wolves affiliating themselves with prehistoric bands of human beings in order to scavenge for food.[51] But could the change (also) have reflected the animal's own intrinsic evolutionary process, as topo-dimensionally conceived? According to the analysis offered in *Topologies of the Flesh*, the Moebial *anima* indeed evolved from a being of primal ferocity to one of loving affiliation, a transpersonal process whose primary site of action was the 2 + 1-dimensional world. But I won't attempt to unpack the details of my earlier analysis here. For present purposes let's just say that—if we assume the validity of the topo-dimensional account, we may surmise that the third Kleinian Proprioception returning us to a more primordial stage of human development is coupled with a Moebial Proprioception returning the *anima* to a more primitive stage in its own development, a transition entailing a backward move in which the loving affiliation that had evolved through animal individuation reverts reflectively to primal ferocity. This proposition does presuppose that animals are capable of reflective consciousness, but the order of reflectivity involved here would be a world apart from that of human beings, for it would be associated with the 2 + 1-dimensional world of the *anima*.

What of Baby Stevie's other nonhuman companion? I propose that this being is not of the animal world but of the *vegetable*, the realm of primordial impulse and instinct, sensation and sexuality. In the third round of the *coniunctio*, Kleinian and Moebial Proprioceptions are thus synchronized with a lemniscatory Proprioception that unleashes the forces of vegetative sensuality. Put in terms of self-signification, the Kleinian Proprioception of Baby Stevie's intuited "I" is coupled with the Moebial Proprioception of his animal alter-"I," and with the lemniscatory Proprioception of his vegetable alter-"I." I submit, moreover, that, while the Kleinian aspect of the triple Proprioception entails the cognitive self-intuition of the primitive human brain, the Moebial aspect involves an *emotional* self-intuition of the wild animal *heart*, and the lemniscatory aspect a *sensuous* self-intuition of the vegetable "groin" (plants do not literally have groins, of course, but they do possess organs of reproduction: their flowers [in the case of flowering plants]).

What we are seeing is that the magical *cogito*'s noncognitive companions also function Proprioceptively through the medium of intuition. So, whereas the Moebial *anima* expresses itself via heartfelt vocalization in its first Proprioception, the medium for its second Proprioception is an emotional form of intuition. In a similar fashion, Proprioceptive awareness of the primal sensuality of the lemniscatory *vegeta* is mediated by intuition of a sensuous kind. (The several types of intuition and their roles in development are elucidated in *Topologies*.) To gain further insight into the arche-erotic realm brought into play with the participation of the 1 + 1-dimensional lemniscate, let us resume our exploration of shamanism.

We know from the last section and previous chapter that the goal of shamanism is an ecstatic return to an earlier "paradisiac epoch."[52] And we have found that, in this journey, plants play a central role. Shamanic practice includes both the eating of plants and their burning (so as to produce a pungent odor); both can have an intoxicating effect on the practitioner, one that generates "magical heat," facilitating entry into a state of ecstasy. Under the narcotic influence of the plants, "the narcotized person 'grows hot'; narcotic intoxication is 'burning.' . . . Often the shamanic ecstasy is not attained until after the shaman is 'heated.'"[53]

Would we not expect that the state of "boiling ecstasy" attained by the shaman should be accompanied by the arousal of sexual desire? Eliade confirms this in detail.[54] In one account, a Siberian shaman tells of how he is possessed by erotic spirits: "whether big or small, they penetrate me, as smoke or vapour would."[55] Is it possible that this "smoke or vapour" that stirred the shaman's loins had a *vegetable* origin; that it was occasioned by the burning of potent incense? Typically, shamans do become *incensed* in the course of their ecstatic journeys. They go into violent paroxysms reminiscent of epileptic convulsions.[56] In his frenzied agitation, the shaman may submit himself to torments: "He gashes himself with knives, touches white-hot iron, swallows burning coals."[57] In this heated condition, the shaman literally fumes.

Joseph Shipley establishes some interesting etymological links in this regard. The Indo-European root "*dheu I*" is generally related to "smoke; dust . . . strong smell," from which derives the Greek "*thuos*: incense. *Thyme*: to cause to smoke; used in sacrifices."[58] The same root gives the Latin "*fume, fumigate; perfume . . . fury*, from the excesses after inhaling incense at sacrificial festivals."[59] What is the upshot of all this? It is that, beyond the shaman's entry into the sonorous world of animal feeling, he or she enters an even denser, more primordial vegetable domain of powerful scents and sensuality, a realm of primal fury not unlike the sulfurous underworld so frowned upon in the patriarchal scriptures.

But the shaman does not merely descend. Her journeys to the "nether world" are not mere bouts of regression to a time when the immature *cogito* was overpowered by tempestuous nonhuman forces. There is a "method to her madness." She is deliberately seeking enlightenment, and Eliade emphatically distinguishes shamanism from psychopathy. He notes that "for all their apparent likeness to epileptics and hysterics," shamans "control their ecstatic movements."[60] To take one example, at the same time that a certain Yakut shaman "'gashed himself with a knife, swallowed sticks, [and] ate burning coals,'" he "'bubbled over with intelligence and vitality.'"[61] In general, the shaman demonstrates an "astonishing capacity to control . . . ecstatic movements," and, "'intellectually, he is often above his milieu.'"[62] It is because the shaman maintains the light of intellect, preserves reflective consciousness during her excursions, that her descent is at once an *ascent*, as we saw earlier in accounts of "magical flights" to "Heaven" enacted through the vegetable magic involved in drumming and climbing the Cosmic Tree.

What bearing does all this have on the matter of self-signification? Evidently, the self-signification of the text now at hand entails a shamanesque transformation in which we "turn up the heat." The "regression in the service of the text" that brings Baby Stevie to presence will indeed be incendiary.[63] For the infant will come forth with all the primal energy and boiling sensuality stemming from the *vegeta*'s influence upon him. To further facilitate Baby Stevie's appearance, pungent incense may be burned that helps trigger the connection with early childhood.[64] Yet the "descent" will surely also be an "ascent," for the presencing of Baby Stevie will be no *mere* regression, since Steven will be there with his reflective intelligence to assist in containing the process. Thus Baby Stevie, come of age and now functioning in the manner of a shaman, will ecstatically enact the magical Proprioception, taking back in the cognitive intuition of himself upon which his text is based, and, in so doing, directly apprehending the sensuous reptilian brain. In the course of this experience, the initiate will certainly not encounter an externally projected god (as the Altaic shaman encountered "Bai Ulgän"). It is *himself* that he will meet, a universal Kleinian Self whose subtle body is sealed for the third time.

Of course, the Kleinian Proprioception will not be the only ecstasy to occur here. For when Baby Stevie appears on the scene, his animal and vegetable alter egos will accompany him. The feral quality of the magical child reflects the influence of the wild *anima*; the infant's potent sensuality (his "polymorphous perversity," as Freud would say) derives from his relation to the *vegeta*. In commencing his Proprioception, Baby Stevie-become-shaman unleashes his nonhuman companions, who are now free to perform Proprioceptions of their own. Thus it is that the *anima*, at present more "heated" than before, can enact its second Proprioception, the self-signification of the text-as-emotional-intuition that entails an immediate apprehension of the primitive animal heart. In the course of this rapture, the generic body of the Moebial Self is hermetically sealed for the second time. Beyond that, the *vegeta* carries out a Proprioception of the text-as-sensuous-intuition that establishes a direct link to the vegetative organs of reproduction and hermetically seals the universal body of the lemniscatory Self. In the threefold Proprioception that transpires, topo-dimensional Selves turn backwards in synchrony as they individuate.[65]

Yet here Steven sits, staring once more at the blinking cursor. Though he has written of the primal forces that drive Baby Stevie, he does not feel them in a palpable way. Steven is reluctant to undergo the regression necessary for the self-signification of the magical text. An alchemical ordeal must be faced, a shamanic rite that reaches its climax in the all-consuming flames. But facing the fire is unthinkable; it is more than Steven can bear. Mindful of this, his thoughts now return to a dream recounted above . . .

Dream of December 10, 1987, "The Woman Inside/The Woman Outside" (Reprise):

A woman is imprisoned behind a brick wall. She is being held for the performance of a dreadful ritual: she is to be burned alive. I am in a state of terrible anguish about the unspeakable agony the victim is to undergo. More than anguish. The thought of someone going through this hellish process, the thought of what she is to experience, is something just too palpably awful for me to bear.

An older woman stands outside the brick wall and prepares to give the command to begin the ceremony. The woman outside is adamant, solemnly insistent that the ritual be carried out. Her heart is hardened to the ordeal the woman inside is about to endure. I find this intolerable.

What I feel in this dream is a sense of astonishment, complete disbelief that any person could undergo such suffering, and that another could have the heartlessness to inflict it. And I am overwhelmed by an uncanny sense of the impossible: how can an event so totally unthinkable actually be coming to pass? But the dream also has an undertone of solemn dignity that accompanies the incomprehensible cruelty. A primal ritual is being enacted that is beyond my capacity to accept or understand.

The ritual must be enacted. The unthinkable must come to pass. The flames will consume the one who stands present for transformation. But if the process is *alchemical,* must the body consumed not also be reborn? Mustn't the finite body that succumbs to the fire rise phoenix-like from its own ashes and be transformed into an *infinite* body, a subtle body? The secret of death and rebirth lies in the way the alchemical process is *contained.*

Long have I pondered the question of alchemical containment. The issue has even infiltrated my dreams.[D1] I believe it helps to regard the vessel that holds the transformative flames as a kind of *crucible,* a container used for heating substances to high temperatures without itself melting down. A properly *alchemical* crucible needs to be configured in a *self*-containing way. It must *be* what it contains. The body placed in the crucible will indeed be consumed by the flames, but the crucible itself will not be consumed, and, since this vessel will in fact be a *vas Hermeticum* (fig. 6.3), the finite body contained will be transmuted into the uroborically infinite body of the container. It is in this immolation of the particular body that the magical Self's subtle body will be fashioned.

I named my 1987 dream of ritual immolation "The Woman Inside/ The Woman Outside." Until now, the dream has been as incomprehensible to me as it has been disturbing, but I think I can finally appreciate its redemptive character: the wall containing the woman inside and separating her from the woman outside is configured uroborically. The cre(m)ation chamber is a Kleinian crucible. In this paradoxical enclosure, the woman within being sacrificed and the woman outside presiding are *one woman*—she who sacrifices herself to be reborn (fig. 6.4).

Of course, to complete the formation of the magical subtle body, to hermetically seal the crucible and bring the third *coniunctio* to fruition, Proprioceptive awareness must be brought to bear. It is through this backward movement of intuitive self-consciousness into the reptilian chamber of the brain that we reenter the fiery realm of the primal infant, perishing in the flames only to rise again in the manner of the legendary phoenix.

Figure 6.3 Uroboros and Hermetic vessel, from *Aurora Consurgens*, fifteenth-century treatise on alchemy (reprise of fig. 2.2).

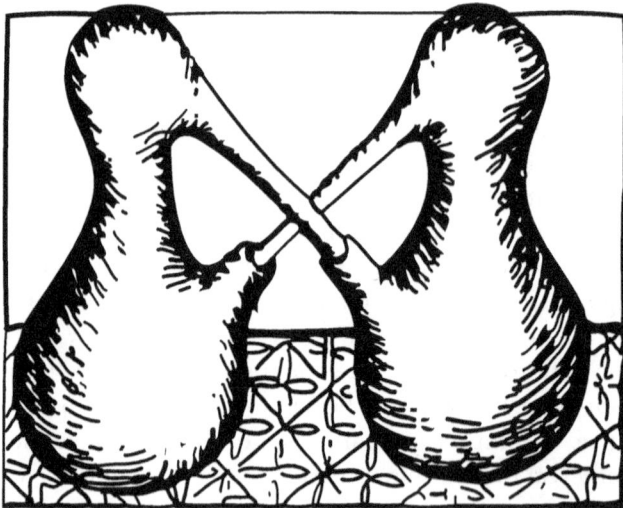

Figure 6.4 Double pelican as "womanly" crucible. The double pelican is an alchemical apparatus discussed in chapter 2 and compared with the Klein bottle in chapter 3. We may imagine this strange-looking vessel as a crucible, with each of its chambers containing the other, being simultaneously inside and outside of one another—as are the "woman inside" and "woman outside" of my dream.

DREAM JOURNAL

D1

Dream of December 11, 2011:

A theoretical question comes up about containment. *This is a big problem for alchemy, since the heat generated in the Hermetic procedure is far too great for the process to be contained. But I think I have the answer and start writing it down on a large index card. Using both sides of the card, I'm writing and writing to make sure I get it all in.*

The solution has to do with the fact that, while containment is definitely a challenge given the extreme heat that is produced, if the container has the properties of a Klein bottle, *the contents would be enclosed in such a way that, paradoxically, they also would be allowed to burst out!*

CHAPTER SEVEN

Final Stage of Alchemical Conjunction

Cogito, Anima, Vegeta, and *Minera*

Would it be entirely accurate to say that we have now completed the first three stages of the Hermetic opus and are ready to proceed to the fourth and final stage? Not entirely. That is because the stages are not so much sequenced in linear fashion with clear-cut points of termination as they are *ongoing dynamic processes* (perhaps we should be speaking of *stagings* rather than "stages"). Accordingly, the *coniunctios* of previous chapters have not been brought to simple closure and left behind; rather, having been initiated, they continue to be enacted, even as we enter the final stage of individuation. The *unio mentalis* of chapter 4 is still unfolding via the Proprioception of Steven, the withdrawal of his Projection by the mental-rational Self. Stevie, the dream child who first appeared in chapter 5, must go on making his presence felt so that the world of mythic consciousness can be further plumbed in the Proprioceptive way. And Baby Stevie, the infant of chapter 6, must continue to be unveiled, leading to the Proprioception of magical mentation. These acts of Proprioception are of course all indigenous to the human realm of cognition, but they are synchronized with co-Proprioceptions carried out by Stevie's and Baby Stevie's nonhuman companions, the feeling *anima* and sensuous *vegeta*.

And now a new conjunction is at hand. In it, the realms of human cognition, animal feeling, and vegetal sensuality are to be brought into harmony with the densest and most concretely embodied realm: that of *mineral intuition*. Topo-dimensionally speaking, this *coniunctio* entails synchronizing the 3 + 1-dimensional Kleinian vessel, 2 + 1-dimensional Moebial vessel, and 1 + 1-dimensional lemniscatory vessel with the zero-dimensional vessel corresponding to the remaining member of our bisection series.

We know that the Klein bottle, Moebius strip, and lemniscate constitute a series of nested topological forms (fig. 5.4). Bisecting the Klein bottle produces the Moebius, and bisecting the Moebius gives the lemniscate. But one more bisection is required to complete the series. Upon cutting the lemniscate, the surface neither retains its integrity nor simply falls into separate pieces. Instead, the single surface is transformed into two interlocking surfaces, each of which is itself lemniscatory. The transformation brought about by this bisection is clearly the last one of any consequence, since additional bisections, being bisections of lemniscates, can only produce the same results: interlocking lemniscates. The bisection series is thus exhausted when we obtain interlocking lemniscates. Let us call this figure the *sub-lemniscate* (fig. 7.1).

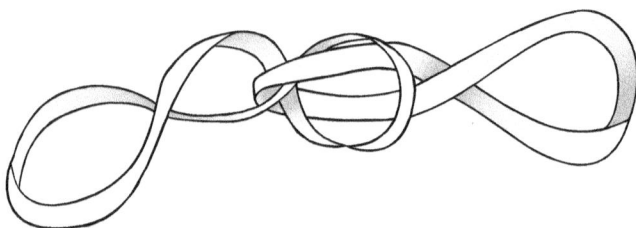

Figure 7.1 Sub-lemniscate.

Like its topological counterparts, the sub-lemniscate, when conventionally observed, appears as but a two-dimensional object in three-dimensional space. In previous chapters, we discovered the deeper meanings of the Moebius and lemniscate. Beneath their appearance as mere objects in three-dimensional space, they constitute lower-dimensional worlds unto themselves: those of the 2 + 1-dimensional *anima* and 1 + 1-dimensional *vegeta*, respectively. The sub-lemniscate also possesses a deeper significance of this kind. At bottom, it corresponds to the zero-dimensional realm of the *minera*. We are about to see that the sub-lemniscatory sphere is unique among worlds, and uniquely mysterious.

THE ENIGMA OF ZERO-DIMENSIONALITY

Note that the dimension number assigned to the sub-lemniscatory world departs from the $n + 1$-dimensional formula used for the higher-dimensional realms. Said formula reflects the idea that each of the higher-dimensional worlds constitutes an order of space and time (which I have shown to be associated with object and subject, respectively). An implication found in previous chapters is that space-time differentiation increases as we ascend from lower- to higher-dimensional realms. In chapter 5, for example, I noted the more intimate overlapping of space and time in the lower-dimensional Moebial world than in the Kleinian, and in chapter 6, I mentioned the near inseparability of space and time in the lemniscatory domain. But we can still say that a space-time distinction is possible in the latter sphere, whereas, in the *sub*-lemniscatory world, such a positive distinction can no longer be made. It is for this reason that the $n + 1$-dimensional form of notation cannot be applied to the world corresponding to the sub-lemniscate. In this zero-dimensional realm, we have neither space nor time, object nor subject, self nor other. Then neither can we have any self-development here, any Projection of finite selves or egos by an infinite Self seeking to individuate, nor any subsequent Proprioception of that Self to complete the individuation process. So, if the zero-dimensional state of affairs is spaceless and timeless, objectless and subjectless; if the sub-lemniscatory world is the only one that does not constitute an order of Self—can we say what it does constitute? I suggest it corresponds to the *unus mundus* adumbrated in chapters 2 and 4.

To reiterate: According to Jung (citing Dorn), the *unus mundus* is the "One World," the "potential world of the first day of creation, when nothing was yet 'in actu, divided into two and many, but was still one.'"[1] Can anything be done to conceptualize this unitary world in positive terms? Jung comments that "there is little or no hope that the unitary Being can ever be conceived, since our powers of thought and language permit only of antinomian statements" with respect to it.[2] Nevertheless, "we do know beyond all doubt, that empirical reality has [such] a transcendental background."[3] When Jung likens the "antinomian" encounter with the *unus mundus* to the "ineffable . . . experience of *satori* in Zen," it appears that only a logic of Zen can do justice to this realm I am associating with zero-dimensionality.[4] Consequently, if we follow such a logic, we cannot allow the implication, as Jung did, that the *unus mundus* is "one" *as opposed to* being "divided into two and many." We must say instead that it entails *neither* unity *nor* diversity.

The Kyoto philosopher Tanabe Hajime linked the principle of neither/nor and Zen with the following words: the "logic of Zen . . . is [that of] *neither/nor*—as in the phrase 'Neither do I say that it is life, nor do I say that it is death.'"[5] According to Tanabe, it is the double negation that permits us to surpass the relative concept of "nothing" and arrive at the notion of *absolute nothingness*.[6] Here negation is not defined strictly in relation to affirmation, simply constituting its lack; rather, "nothingness means transformation."[7]

Tanabe strongly emphasizes the *processual* character of nothingness. Throughout his principal work, *Philosophy as Metanoetics*, he associates nothingness with action, transformation, mediation. Absolute nothingness is grasped as "absolute mediation" or "absolute transformation," as "pure movement" or as an absolute "flux of activity . . . an incessant conversion in which being is transformed into nothingness and nothingness into being."[8] Can we not read "being" as "life" and "nothingness" as "death"? Tanabe himself implicitly does as much in critiquing Heidegger's concept of death: it "is a death interpreted entirely from the standpoint of life, a nothingness interpreted from the standpoint of being."[9] What Tanabe apparently wants to bring out is the dynamic interpenetration of life and death. Thus a prominent theme for Tanabe is that of *death-and-resurrection*, with the hyphens indicating no categorical separation of life and death: "one is restored to life only in the form of one who is dead."[10]

Of course, absolute nothingness "itself" cannot die and be reborn in this paradoxical way, since zero-dimensional action as such *has no self*; it is utterly selfless, devoid of life or being in its own right. Rather than engaging in its own individuation process, the zero-dimensional milieu—as "absolute mediation"—mediates the individuation of the higher-dimensional Selves. It is *their* death-and-resurrection that is facilitated in the selfless matrix. Recalling von Franz's identification of the *unus mundus* as the "realm of the dead" (see chapter 2), we may now add that this zero-dimensional matrix is one of *absolute* death, in keeping with Tanabe's characterization of it as absolute nothingness. By the same token, we may speak topologically of its absolute "holeness."

We know in general that the subtle body enacts a dialectic involving the differentiation and interpenetration of life and death, being and nothingness, continuity and discontinuity, space and time, object and subject, outer and inner, plenum and void. Expressed topologically, this dialectic operates in the relationship of the Klein bottle's continuous aspect to its hole. Descending into lower dimensionalities, said dialectic of whole and hole remains in evidence, but in a weaker form. For, as I noted above, space and time—associable with whole (continuity) and hole (discontinuity), respectively—are not as sharply differentiated in the 2 + 1-dimensional Moebial world as they are in the 3 + 1-dimensional Kleinian world, and even less differentiated going down another dimension into the 1 + 1-dimensional lemniscatory world. This progressive weakening in the dialectic of (w)holeness with topo-dimensional descent from Kleinian *cogito* to Moebial *anima* to lemniscatory *vegeta* reaches its climax in the world of the sub-lemniscatory *minera*, where whole and hole are utterly undifferentiated. Applying Zen logic to this domain, we may say that it is *neither* whole *nor* hole. Just as Tanabe characterized the neither/nor of being and (relative) nothingness as *absolute* nothingness, we may topologically describe the zero-dimensional realm as involving absolute *holeness*.

Note the etymological relationship between the words *hole* and *hell*: both derive from the Indo-European *kel*, meaning "to cover," "conceal." The netherworld or "realm of the dead" is commonly depicted as a hollow or hole deep in the bowels of the earth. When we characterize this subterranean sphere as constituting *absolute* holeness, its non-finite nature is brought out: it is not merely a circumscribed hollow beneath the ground but a world unto itself—the zero-dimensional world of the *minera*.

It is minerality that predominates beneath the earth. Of course, the ordinary *human* experience of the mineral kingdom is a far cry from the zero-dimensional sphere as such. From the standpoint of the 3 + 1-dimensional *cogito*, the world of minerals consists merely of discrete inanimate matter: crystals, metals, rocks, the stuff of inorganic chemistry and physics. Such an objectifying, deadening representation of the mineral domain hardly does justice to minerality per se. To apprehend the zero-dimensional *minera* in a deeper way, we turn once more to the concrete intuitions of alchemy, that ancient source of chemistry and physics.

In alchemical terms, the mineral kingdom gains ultimate symbolic expression in the *lapis philosophorum* or *Philosopher's Stone* that we explored in chapter 2. This mysterious substance certainly does not consist of matter that is merely inert, objectively delineable. According to Jung, the lapis was:

> the *filius macrocosmi* as opposed to the "son of man," who was the *filius microcosmi*. This image of the "Son of the Great World" tells us from what source it was derived: it came not from the conscious mind of the individual man, but from those border regions of the psyche that open out into the mysteries of cosmic matter.[11]

Later in the same volume, Jung says that "alchemical projections . . . point back to something primeval, to the apparently hopelessly static, eternal sway of matter. . . . They show us, as the redemptive goal of our active, desirous life, a symbol of the inorganic—the stone—something that does not live but merely exists or 'becomes.'"[12] Correlated with the stone is the unconscious psyche, which "is refractory like matter, mysterious and elusive, and obeys laws which are . . . nonhuman."[13]

In *Psychology and Alchemy*, Jung notes:

> The Christian projection acts upon the unknown in man. . . .
> The pagan projection, on the other hand, goes beyond man and
> acts upon the unknown in the material world, the unknown substance which, like the chosen man, is somehow filled with God.
> And just as, in Christianity, the Godhead conceals itself in the
> man of low degree, so in the "philosophy" [i.e., in alchemy] it
> hides in the uncomely stone. In the Christian projection the *descensus spiritus sancti* stops at the *living body* of the Chosen
> One, who is at once very man and very God, whereas in alchemy the descent goes right down into the darkness of inanimate
> matter.[14]

Although the alchemist's projection may make it appear that matter requires the descent of spirit in order to be animated, Jung proceeds to relate the "inanimate matter" of the apparently "uncomely stone" to the *anima mundi*, which we know is a fundamental principle of animation in its own right.[15] What this suggests is that the *anima mundi* is not only associated with the animal and vegetable domains of nature, as established in previous chapters, but also, and most primordially, with the "inorganic" mineral realm. It would seem then that *all* of nature's dimensions entail dynamic activity. However, this conclusion is somewhat blunted by the logic of absolute negativity that governs the zero-dimensional sphere. While the *minera* surely is not inert, neither can we simply describe it as active. For when it comes to such terms of opposition, in the sphere of absolute nothingness the rule is *neither/nor*!

Now, despite the enigmatic negativity of the mineral world, I intimated at the outset of this chapter that we can associate the zero-dimensional domain with a function of the psyche, namely, *intuition*. This follows from what we've already established: each of the three higher-dimensional spheres is principally governed by a different psychic function, as specified by Jung: the human world is ruled predominantly by thinking, the animal realm by feeling, and the vegetable sphere by sensation or sensuality. Evidently, the *fourth* basic function set forth by Jung serves as the governing principle for the fourth realm of nature: intuition prevails in the mineral world. This is consistent with the above-noted link between stony minerality and the unconscious psyche, which "is refractory like matter, mysterious and elusive, and obeys laws which are ...nonhuman." For we know from chapter 4 that intuition is the psychic function that operates chiefly under the auspices of the unconscious. In that chapter, we also explored the developmental implications of Jung's analysis of psychic functioning—how the functions lend themselves to alignment with the alchemical stages, with each function corresponding to a different stage of conjunction. It is the function of intuition that is aligned with the fourth and final *coniunctio* now at hand, that in which we are to descend into the zero-dimensional "realm of the dead" to seal the alchemical vessel in stone.

In chapter 6, we found intuition playing a role in the *third* conjunction. Here it serves as the medium of self-signification for the three individuating higher-dimensional Selves. What we need to do now, in enacting the fourth *coniunctio*, is trace intuition back down to its source in the zero-dimensional *unus mundus*. At bottom, intuition is indeed the most mysterious of the four functions, the most nebulous and primordial function, the one Jung associated with the "lure of possibilities."[16] Already in chapter 4, I related these "possibilities" to the *unus mundus*, to the "potential world of the first day of creation." The possibilities in question are essentially those of the higher-dimensional Selves, their potentials for individuation. Embedded in the zero-dimensional matrix, the Selves are no more than embryos that are not yet capable of functioning on their own, not even intuitively; here they are germinally realized by "Great Mother's" intuition of them. Being utterly Selfless, the *minera* does not function intuitively for itself, but only for the Selves that incubate in its midst. The archaic structure of human consciousness of which Gebser speaks is one such embryonic Self.

ARCHAIC CONSCIOUSNESS

We have discovered that the individuation of the cognitive Self reaches fruition with the fourth sealing of the Kleinian vessel. This final *coniunctio* calls for lifting the repression of the most primitive structure of consciousness described by Gebser: the *archaic*. Associating this embryonic form of awareness with a state of "deep sleep" (deeper than the magical state of sleep), Gebser describes it as involving "the nondifferentiation, indeed the non-distinguishability of ... man from world and universe—a non-awakeness by virtue of which he is still unquestionably part of the whole . . . an unconcerned accord, a consequent full identity between inner and outer . . . the perfect identity of man and universe."[17] We may wonder, however, whether "whole" and "perfect identity" are appropriate terms for this initial state of affairs. If the undifferentiated situation prevailing at the outset of development is understood in terms of zero-dimensionality, the logic of *neither/nor* should obtain: archaic consciousness should neither stand apart from, nor be simply immersed in or identified with, a seamless whole—for there would be no such whole, any more than there would be separate parts. As an alternative, I am proposing that we regard the 3 + 1-dimensional archaic structure as an embryonic potential embedded—not within a whole, but in the zero-dimensional matrix of absolute holeness. Let us consider the meaning of "embryonic potential" in this context.

When we apply the neither/nor of Zen to the "potentiality" of the embryonic structure, we realize that the bracketed word cannot signify an *Aristotelian* potential. Though "potential" surely is inherent in the embryo, this can neither mean nascent positive being nor the mere negation of being. Such an embryo is not a pre-formed, miniature version of the mature being, one that is "in there" already and just waiting to come out. Nor is said embryo a *mere* negativity, a state of simple and total formlessness from which any arbitrary form could be created, the creation having to be ex nihilo. In the embryo then, we have a "being" that is curiously poised between being and nonbeing, between actuality and mere nonexistence. We may regard such a potentiality as an *implication*, in the sense articulated by philosopher/psychologist Eugene Gendlin.

In his essay "Thinking beyond Patterns," Gendlin offers the example of a poet searching for just the right words to effectively express the next line of a poem. The poet appeals to a place in her body that Gendlin calls a "blank" or a "slot," signified by "......" In the ......, the next line of the poem is *implied*: "This ...... demands and implies a new phrase that has not yet come."[18] "Yes, the next line is *implied*," says Gendlin, "although it does not exist and never has."[19] By way of explaining further, Gendlin distinguishes the conventional meaning of "implicit" from his own:

> What we have once thought explicitly can become implicit when we stop thinking about it. Much previous thought is also implicit; it has been built into our situations and our lives although we have never thought it explicitly ourselves. But in our instance

(of a poet), what *was* implicit can be new to the world. Let it stand that something quite new can have been implicit.[20]

For our purposes, Gendlin's idea grants the possibility that something that does not exist in any positive fashion nevertheless cannot simply be said to not exist at all. The archaic structure of consciousness can be regarded as existing in this Zen-like "negative" manner, as an *implication*. At one point Gendlin associates the notion of "implying" with being "pregnant."[21] This term accords well with our portrayal of archaic consciousness as embryonic. As the poet is pregnant with the next line of the poem, so the selfless zero-dimensional *minera* is pregnant with the cognitive Self, the embryo taking the form of archaic consciousness.

The deep significance of the embryo is adumbrated in Jung's *Mysterium Coniunctionis*. Here it finds expression as the *point* or *scintilla*. Regarding the former, Jung cites the speculations of alchemist John Dee:

> "It is not unreasonable to suppose, that by the four straight lines which run in opposite directions from a single, individual point, the mystery of the four elements is indicated. . . . Things and beings have their first origin in the point and the monad." The centre of nature is "the point originated by God," the "sun point" in the egg. This, a commentary on the *Turba* says, is the "germ of the egg in the yolk." Out of this little point, says Dorn . . . , the wisdom of God made with the creative Word the "huge machine" of the world.[22]

Jung goes on to say:

> The point is identical with the . . . scintilla, the "little soul-spark" of Meister Eckhart. . . . Hippolytus says that in the doctrine of the Sethians the darkness "held the brightness and the spark of light in thrall," and that this "smallest of sparks" was finely mingled in the dark waters below. Simon Magnus likewise teaches that in semen and milk there is a very small spark which "increases and becomes a power boundless and immutable."[23]

"Alchemy, too," says Jung, "has its doctrine of the scintilla. In the first place it is the fiery centre of the earth, where the four elements 'project their seed in ceaseless movement. For all things have their origin in this source, and nothing in the whole world is born save from this source.'"[24]

Jung then associates the scintillae with "fishes' eyes" ("*oculi piscium*"). "The eyes indicate that the lapis [stone] is in the process of evolution and grows from these ubiquitous eyes."[25] That the seed-like scintillae are embedded in the original stone is brought out in saying, "'Upon one stone there are seven [fishes'] eyes.'"[26] Next, making the connection with consciousness, Jung comments that "the fishes' eyes are tiny soul-sparks. . . . [that] correspond to the particles of light imprisoned in the dark Physis. . . . The eye, like the sun, is a symbol as well as an allegory of consciousness. . . . 'The eye is . . . clarity of intellect.'"[27] Finally: "'The Son of the Great World [i.e., the lapis] . . . is filled, animated and impregnated . . . with a fiery spark [scintilla] of *Ruach Elohim* [God]'. . . . The 'fiery sparks of the World-Soul [*anima mundi*]' were already in the chaos, the prima materia, at the beginning of the world."[28]

How may we translate all this into the terms of our present discourse? In the seed-like scintillae or "eyes" that dimly illuminate the darkness of the zero-dimensional *unus mundus*, we have the higher-dimensional "*I*'s," the embryos of ego consciousness associated with our several topo-dimensional bodies. These germinal elements are already present in the original chaos of the sub-lemniscatory *minera* and correspond to the *potentialities* of "the potential world of the first day of creation." In this scheme, the embryo of 3 + 1-dimensional human consciousness is the archaic structure that Gebser wrote about.

Archaic consciousness indeed has egoic potential, in contrast to the zero-dimensional *minera* per se, which has none. But we have to keep in mind the unique negativity of the archaic. This is made clear in the difference between the embryonic ego of the archaic structure and the ego of the magical structure discussed in the previous chapter. However inchoate the latter may be, the magical human being does possess a modicum of ego in the positive sense of the word, whereas the archaic ego is but an *implication*. Such an ego cannot be said to exist in any positive manner, though neither can we say that it simply does not exist! A similar conclusion can be reached with respect to the space-time structure of the archaic. Unlike the zero-dimensional matrix, 3 + 1-dimensional archaic consciousness is not merely spaceless and timeless. Yet this embryonic order of the human world does not possess the positive space-time structure of any of its more well-developed counterparts (the magical, mythic, or mental-rational).

Is archaic consciousness correlated with any particular organ of the body? In the last chapter, we took up Gebser's issue of the organ centers to which the several structures of consciousness are related. For Gebser, the brain is associated with the mental-rational structure, the heart with the mythic structure, and the intestines with the magical structure. He specified no organ correlate for archaic consciousness. As a first step, I proposed amending Gebser's account to align the magical structure with the reproductive organs, leaving the archaic structure to be associated with the intestines. Then, departing further from Gebser, I qualified this analysis by suggesting that, while each of the older structures of consciousness (mythic, magical, archaic) is indeed strongly influenced by a sub-cranial organ center of the body (heart, reproductive organs, and intestines, respectively), the *primary* organ correlates of these cognitive structures are sub-cortical centers of the *brain* (paleo-mammalian, reptilian, and neural chassis, respectively). Let me now say a little more about the neural chassis.

Although Paul MacLean postulated that the human brain is triune (fig. 5.9), he acknowledged that there actually exists a "fourth brain" serving to ground the other three: the *neural chassis*. According to MacLean, the basic neural machinery required for self-preservation and the preservation of the species is built into the lower brain stem and spinal cord. This neural chassis is comparable to the chassis of an automobile. If, by itself, "the neural chassis . . . [is] likened to a vehicle without a driver," evidently there evolved three different drivers for the same vehicle, "each radically different in structure, chemistry, and organization."[29] The "three different drivers" are of course the neo- and paleo-mammalian brains, and the reptilian brain. In considering MacLean's model, researcher Erich Jantsch observes that the "neural chassis . . . is older than the three drivers and may be traced back to an early phase of the appearance of multicellular organisms."[30] If this primeval brain "were not driven by three 'drivers,'" comments Jantsch, it "would resemble an empty and unguided vehicle."[31] Phylo-functionally, the neural chassis is the area of the brain associated with the archaic thinking of the *cogito*.

What about the influence of the *intestines* on archaic consciousness? We have established that mythic consciousness is primarily associated with the old mammalian brain but strongly affected by the animal heart (with which the human heart resonates) and magical consciousness is chiefly linked to the so-called reptilian brain but influenced by the vegetal reproductive system (to which the human genitalia are attuned; we know, for example, that the scent of certain plants can trigger human sexual arousal). It seems to follow then that, while the archaic structure would be primarily tied to the neural chassis, it would also feel the effects of an intestinal organ center that would be related to the mineral world. And yet, though animals indeed have hearts and vegetables have reproductive organs, does it make any sense to speak of the intestines of minerals?

To be sure, dead matter could not literally have intestines! Yet the mineral world in fact is *not* simply a realm of dead matter, as the post-Renaissance objectivist would have it—not when it is understood alchemically. Inasmuch as the zero-dimensional *minera* is actually the creative matrix for all life—the "potential world of the first day of creation"—it is the germinal realm in which organ systems originate. But why specifically associate the *minera* with the *digestive* system, with the intestines and alimentary canal?

The most basic fact of all life is that it must eat—from the elaborately specialized digestive processes of human beings with their alimentary canals and intestinal tracts, to the functioning of one-celled organisms that simply *are* digestive systems if nothing else, ingesting nutrients and expelling wastes to maintain their identities as bounded entities, proto-egos. Viewed alchemically, what distinguishes the mineral world is not that it consists of lifeless matter devoid of alimentation but that, on the contrary, more than just constituting the ultimate source of the most basic organ system, there is a direct sense in which the *minera* per se can be said to be alimentary. This was intimated in chapter 2 when I noted alchemy's identification of its long-sought "mineral," its lapis or Philosopher's Stone, with its primary symbol: the uroboros (fig. 5.8). For the image of the uroboros or "tail-eater" clearly suggests digestive activity. Further attesting to the mineral nature of the "alimentary uroboros" (as Neumann calls it) is its association with *mercury*, a fundamental element of alchemy, and personified as Mercurius.[32] The close link between the uroboros and Mercurius was established in chapter 2, and we saw that both serve to personify the *unus mundus*, the "one world" that we associate here with the zero-dimensional realm of the *minera*. So the uroboros is not only depicted as an animal, a snake or dragon that digests itself, but also as a mineral. As Jung puts it, "Mercurius . . . is the poisonous dragon and at the same time the heavenly lapis."[33] Indeed, the uroboros is a mercurial shape-shifter par excellence and we know from previous chapters that it can take varying forms. In the next section, we are going to explore its several manifestations in greater detail, whereupon we will discover that the image of the uroboros most appropriately associated with the *minera* and archaic consciousness is not the same image of the tail-eater used in chapter 5 for mythic consciousness.

For his part, Neumann makes much of the "alimentary uroboros." To begin with, he brings out that the intestines and alimentary canal are the most primal regions of bodily functioning:

> Deeper down [than the brain, heart, and sexual organs] lies the psychic plane of intestinal processes of the alimentary tract. . . . For the embryonic ego the nutritional side is the only important factor, and this sphere is still very strongly accentuated for the infantile ego, which regards the maternal uroboros as the source of food and satisfaction. The uroboros is properly called the "tail-eater," and the symbol of the alimentary canal dominates this whole stage. . . . On this level, which is pregenital because sex is not yet operative . . . there is only a stronger that eats and a weaker that is eaten.[34]

Neumann then proceeds to point out the *cosmic* significance of the alimentary uroboros. The "uroboros is coordinated with cosmogony," he says.[35] And in "primitive psychology and mythology the 'alimentary uroboros' is a cosmic quantity."[36] Regarding the uroboros as cosmic, we confirm that its alimentary action indeed encompasses the *mineral* world, not just that of organic life.

Neumann goes even further, noting that, in ancient East Indian mythology, the digestive process takes on an *infinite* aspect:

> The world, according to an early Vedic idea, was created by Prajapati, who is at once life and death—or hunger. It was created in order to be eaten as the sacrifice which he himself offers to himself. This is how the horse sacrifice [of the Upanishads] is interpreted, the horse standing for the universe, like the bull in other cultures: "Whatever he [Prajapati] brought forth, he resolved to eat. Because he eats (*ad*) everything, he is called infinite (*aditi*). Therefore, he who knows the essence of *aditi*, becomes the eater of the world; everything becomes food for him."[37]

Neumann concludes that, in old Hindu religion, "to 'know the essence of *aditi*' is to experience the infinite being of the creator who 'eats' the world he has created."[38] Therefore, in this tradition, digestion seems to possess ontological significance. In the alchemical variation on this view that I am proposing, we do not have a fatherly creator who fashions the world. Instead we have at the outset an infinite world soul (*anima mundi*) or world matrix, the zero-dimensional *prima materia* from which the higher-dimensional worlds are generated. And this uroboric mineral world (a.k.a. the *unus mundus*) prevailing at the very origin is in fact not an ontological realm, not a realm of Being or Self; rather, it is *meontological*, a selfless sphere of absolute nothingness.

No doubt the zero-dimensional alimentary *minera* exerts a potent influence on the 3 + 1-dimensional archaic *cogito*. For, at this point in its development, the human *cogito* is embedded within the mineral matrix as but a seed (a scintilla). Nevertheless, since the brain is the organ native to human cognition, it seems we must still say that archaic cognition is *primarily* associated with the neural chassis of the embryonic brain, rather than the alimentation of the *minera*. But this turns out to be something of a moot point because the neural chassis of the archaic human being is barely differentiable from the alimentary functioning of the *minera* in which said proto-being is implanted. Relevant here is a passage from Neumann suggesting that, at the outset of development, not only is mind inseparable from body, but also, the organs of the body are largely undifferentiated from each other: "Whereas in its later developments centroversion promotes the formation of ego consciousness as its specific organ, in the uroboric phase, when ego consciousness has not yet been differentiated into a separate system, centroversion is still identified with the functioning of the body as a whole and with the unity of its organs."[39] We may say then that, in the primordial matrix of the *minera*, there is little separation of mind and body, of the organs of the body, and of the human body from the rest of nature.

IMAGES OF THE UROBOROS

The symbol of the uroboros could not be more significant for the present work. Having made extensive reference to this symbol throughout the book, I will now attempt to clarify it further, along with its associated structures of consciousness and organ systems.

In portions of the text, I have employed the image of the "tail-eater" somewhat loosely as a gloss for alchemical paradox and topological dialectics in general. Working more specifically with the uroboros in chapter 5, I identified it as the mythic precursor of the Klein bottle, and I used the classical Egyptian image of it as a meditation piece serving to facilitate the self-signification of the mythic text. What we have just seen in the present chapter is the relationship of the uroboros to *archaic* consciousness. So it seems the uroboros can be linked effectively to both mythic and archaic awareness. The same can be said for mental-rational and magical awareness, as implied in previous chapters. All four structures of consciousness can thus be symbolized uroborically. But what we are about to discover is that a different image of the uroboros is called for in each case.

We can already point to two images of the uroboros that each correspond distinctively to a particular structure of consciousness. For, if the
ancient Egyptian precursor of the Klein bottle (see fig. 5.8) is the image
of the uroboros specifically correlated with mythic consciousness, the
Klein bottle itself (figs. 3.6, 3.9, 4.2) may be taken as the modern-day,
mental-rational counterpart of the uroboros (as implied in chapter 3). In
psychiatrist Seymour Boorstein's words, "The Klein bottle is an exqui-
site modern expression of the *uroboros*, for it shows how the inner and
outer worlds connect."[40]
We know that the organ correlate of the Klein bottle is the neo-
mammalian brain, whereas its mythic antecedent is attuned to the paleo-
mammalian brain. But we have also found that another organ plays a
role in mythic experience: the heart. For, while mythic consciousness
is an ancient form of human cognition, it is powerfully influenced by
the heart-centered sphere of animal feeling. Stated in topo-dimensional
terms, although mythic consciousness is primarily linked to the 3 + 1-
dimensional Kleinian world, it bears the stamp of the 2 + 1-
dimensional Moebial world. Consonant with this is the curious fact that
the mythic uroboros possesses a flattened, Moebius-like appearance (fig.
7.2). (But we should not confuse the Moebius strip given to us in three-
dimensional space with the Moebius structure native to the 2 + 1-
dimensional world of the *anima*. The former can be viewed as the "topological shadow" cast by the latter into our higher-dimensional world.)
Interestingly, there is some evidence to suggest that the heart itself is
structured in a uroboric Moebial way.

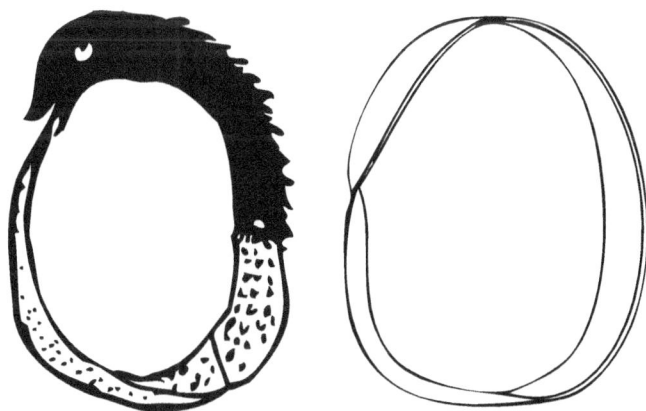

Figure 7.2 Egyptian uroboros (left) and Moebius strip (right).

According to Rapoport, "The actual structure of the [mammalian] heart . . . is that of an iterated Moebius band, as was discovered . . . by F. Torrent-Guasp, in which each band folds another identical band. Furthermore, this development is crucial to the physiology of the heart, which thus functions as a torsioned geometry through vortex motions produced by the cyclical turns of the recursive Moebius bands."[41] A more detailed account of the heart's Moebial anatomy is given by Kocica and colleagues, including an illustration of the helical ventricular myocardial band as modeled by the Moebius strip, and Torrent-Guasp's drawing of the heart (fig. 7.3, left), reminiscent of M. C. Escher's uroboric rendition of the Moebius as three tail-biting fish (fig. 7.3, right).[42]

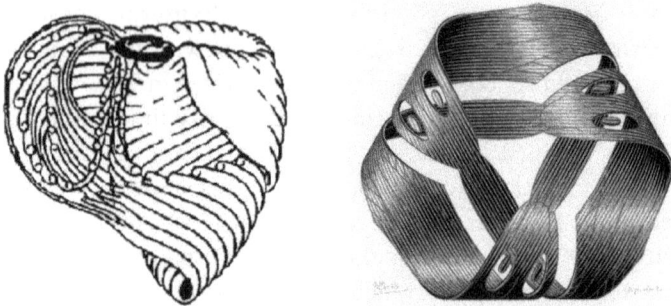

Figure 7.3 Comparison of human heart and Moebius strip.
Left: drawing of the heart, by Francisco Torrent-Guasp (from Mladen J.Kocica et al., "The Helical Ventricular Myocardial Band," *European Journal of Cardio-Thoracic Surgery* 29 [2006]: S29, fig. 9A). Right: M. C. Escher's *Moebius Strip I* © 2014 The M. C. Escher Company—The Netherlands. All rights reserved. www.mcescher.com.

Moving further back in cultural history to magical consciousness, a third image of the uroboros is required, an even more primordial form of the Klein bottle bearing the cast of the next lower member of the topological bisection series. The member I am speaking of is of course the *lemniscate* (fig. 6.2), a figure that has often been depicted as a uroboros (fig. 7.4). Is the lemniscate or figure 8 linked to the sexual organs, as our work with the lemniscatory *vegeta* and magical consciousness would lead us to expect? The connection is made through literature in a powerful work of fiction: James Joyce's *Ulysses*.[43]

Joyce scholar Christine van Boheemen-Saaf speaks of "Joyce's recourse to the lemniscate as a symbol to indicate . . . his notion of the eternal feminine."[44] "The figure 8 . . . symbolizes eternity," and this

Figure 7.4 Uroboric images of the lemniscate.
Left: book illustration from *Azoth*, by Basil Valentine (1659).
Center: carving at gravesite in Erlangen, Germany (courtesy of Janericloebe,
WikImedia.org). Right: tattoo design, by ELJIII (deviantart.com).

shape is brought in "as if it were the figure of femaleness or femininity itself. . . . The figure eight may be associated with the form endearing of the female breasts or the female buttocks."[45] Jacqueline Kay Thomas goes further in her doctoral dissertation on Joyce. She observes that "the infinity sign . . . permeates the text of *Ulysses*" and that "Joyce's subtle assimilation of [his character] Molly to Aphrodite, a goddess whose archaic roots link her to all chthonic earth goddesses, makes the goddess' sexy, unselfconscious narrative presence provide the archaic substratum of Joyce's art."[46] Notable for our purposes, Thomas relates the lemniscate to the uroboros: "Joyce reinstates the infinity sign's most 'primitive' significance when he reconfigures [English essayist Walter] Pater's linear serpentine as a recursive textual *ouroboros.*"[47] So, according to Thomas, Joyce's "larger narrative project" is that "of ironically representing infinity with the invocation of a grounded, physical form (a dance) of the never-ending, serpentine infinity sign."[48]

Granting that Joyce relates the lemniscate to secondary sexual characteristics such as breasts and buttocks, what of the genital organs themselves? Joyce scholar Michael Stanier notes that, not only do "Molly's breasts and buttocks . . . make the lemniscate. . . . [but] this figure is also represented by her genitalia."[49] The point is spelled out by Robert Boyle: "I suspect that Joyce is using the shape and structure of the figure 8, so frequently repeated, for sexual symbolism. . . . If this is so, and if the 8 symbolizes Molly's genital area, as I think it does, then [the character] Bloom, who worships woman particularly in her life-giving and regen-

erative role, is here ritually approaching the source of human life." Boyle
goes on to say that "Molly's vagina" is symbolized by "the upper half of
the eight."[50]

Does the correlation of the lemniscate with the sexual organs go
beyond Joyce's imaginative literary constructions? The lemniscatory shapes of female breasts and buttocks linked to chthonic femininity
have been discovered among ancient relics of prehistoric art. Collectively
known as the "Venus figurines" (fig. 7.5), scores of Paleolithic statuettes
dating back tens of thousands of years have been found across Europe
and in parts of Asia.

Figure 7.5 Venus figurines. Upper left: Venus of Willendorf (courtesy of Matthias Kabel,
Wikimedia.org). Upper right: Venus of Lespugue (courtesy of Locutus Borg, Wikipedia.org).
Below: Venus of Moravany (courtesy of Martin Hlauka/Pescan, Wikipedia.org).

In many of these figures, while arms and legs are attenuated or absent and the head is poorly developed, breasts and buttocks are displayed in an exaggerated manner. And in a number of the artifacts, it is not only the secondary sexual characteristics (including thighs and hips) that are highlighted. The female genital area itself is prominently shown, as in the Venus of Moravany. Yet the vulvae displayed in these images do not seem to have the clearly lemniscatory, figure-8 quality possessed by breasts and buttocks. However, the representation of the vagina featured in some of the Venus figurines can be associated with a geometric structure related to the lemniscate: the *vesica piscis* (fig. 7.6).

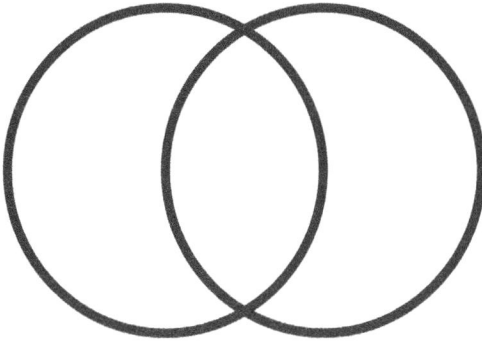

Figure 7.6 The *vesica piscis*.

This Latin term translates as "fish bladder" (one blogger notes that "Vesica Piscis, the 'bladder of the fish,' is probably a euphemism for the 'vagina of the fish'").[51] According to psycho-anthropologist Lawrence Blair: "Once a circle . . . has moved at least one radius distant from itself, it produces an archetypal symbol: the vagina-shaped 'vesica piscis'—the feminine principle of generation."[52] Blair's interpretation accords with that given by artist and writer Buffie Johnson in her *Lady of the Beasts*: "Used in both pagan and Christian religions, the vesica piscis ('vessel [or bladder] of the fish') reflects the idea of the feminine as vessel. Among the pagans, the eating of fish on a certain day represented the deification of the yoni[D1] [the female sexual organ venerated as the origin of life]. . . . In early art and myth the fish glorifies the Great Mother and her womb. . . . [Today] . . . the lozenge or rhomb [which Johnson associates with the *vesica piscis*] is still the visual symbol for vagina."[53]

But what of the *lemniscate*? Exactly how is the figure 8 or sign of infinity related to the *vesica piscis*? The answer lies in working dynamically with the lemniscate. Beginning with the infinity sign, ∞, we can construct a *vesica piscis* by bringing together the lemniscate's two circulations (fig. 7.7) so that the center of each circle lies on the circumference of the other. In fact, anthroposophical geometer Olive Whicher suggests that, by allowing the lemniscate's twin foci to converge, we can generate a whole spectrum of lemniscatory figures, from ∞ → o. In thus merging the lemniscate's centers, "we still have the lemniscatory process *dynamically* present," even though a static view of any figure resulting from the convergence departs from the figure 8 with which we began.[54] To produce the *vesica piscis* in this process, the centers of course do not converge completely to become the *same* center, but come together only to the extent that each circle's center lies on the other's circumference. On this basis, we can indeed conclude that the archetypal symbol of feminine sexuality embodied in the *vesica piscis* is a form of lemniscate.

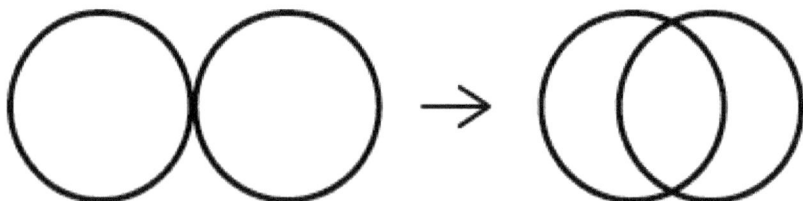

Figure 7.7 Transformation of lemniscate (left) into *vesica piscis* (right).

I should point out that, unlike the image of the Egyptian or Moebial uroboros (figs. 5.8 and 7.2, left), which was used to facilitate the self-signification of mythic consciousness in chapter 5, an image of the lemniscatory uroboros (fig. 7.4) cannot be employed in the self-signification of magical consciousness because the magical self is not primarily imaginal but intuitive, as we saw in the previous chapter. Yet don't magical works of art like the Venus figurines involve visual imagery? To be sure, when we moderns view the statuettes in post-Renaissance perspective as objects in Cartesian space, our experience of them is visual. But we have found that human perception was not dominated by perspectival vision prior to the Renaissance. I suggest that our modern, visually conditioned experience of the Venus figurines departs significantly from the way they were experienced many thousands of years ago when they were first carved. I submit that Paleolithic people generally did not relate to these works so much as objects appearing before their visual gaze, but more as presences with which they were intimately engaged in a sensuous tactile encounter guided by primal intuition (this interpretation seems consistent with the vague depiction or complete absence of eyes and other facial features in the Venus figurines [fig. 7.5], coupled with the exaggerated development of sexual organs). I am suggesting then, that—because visual imagery was at best only weakly developed in a Paleolithic consciousness dominated by nonvisual intuition—the lemniscatory *image* of the uroboros cannot concretely embody the intuitive self of the magical world. But I propose that said image can at least serve us as an archetypal symbol of that world. It is through this image that the pre-imaginal realm of magic makes its presence felt to our visual sensibilities.

Let us return now to the very earliest structure of human consciousness, the archaic. Is there a uroboric symbol of archaic reality that can be associated with the most elemental form of the Klein bottle, that bearing the stamp of the lowest-dimensional member of the bisection series, our sub-lemniscate? I noted at the outset of this chapter that when you bisect the lemniscate, the resulting figure neither remains intact (as happens when you bisect the Moebius to form the lemniscate), nor does it decouple into two pieces that are simply detached from one another. Two lemniscates are indeed produced, but they loop together in an interlocking structure (fig. 7.1). Does there exist a uroboric counterpart of this topological form, a figure that does not consist of a single serpent that swallows itself to form a closed loop as does the lemniscatory uroboros, but a uroboros that is decoupled into two distinct serpents that nonetheless twine together? Indeed such a uroboric figure exists, and it is none other than the *caduceus* (fig. 7.8, left), the figure appearing on the staff of Mercurius (fig. 7.8, right).[55]

In ancient myth, Mercurius (known to the Greeks as Hermes) functions as guide to the dead. Often portrayed as a hermaphrodite, s/he is an emissary to the Underworld granted the ability to pass freely from the gods on high through the terrestrial plane and down into the bowels of the earth (the earth's "intestines"!). This is of course consistent with what I have already proposed: that Mercurius personifies the "realm of the dead," the *unus mundus*, the zero-dimensional nothingness of the sub-lemniscatory *minera*. What I have not brought out until now (and was not even aware of until recently) is that Mercury's staff, with its twining serpents reminiscent of alimentary windings, is the uroboric counterpart of the sub-lemniscate.

This symbol of interweaving snakes mounted on a central staff is actually found more than a millennium earlier than its association with Mercurius. The serpent motif appeared in the third millennium BCE in con-

Figure 7.8 The caduceus (left) and Mercurius with staff (right).

nection with the Babylonian god(dess) Ningishzida (fig. 7.9).[56] Medical historian Walter Friedlander cites archaeologist Arthur Frothingham's view that "the figure of the entwined snakes was itself the god [Ningishzida] and that . . . 'the caduceus is held in the right hand either of the Mother Goddess (Ishtar) or, more rarely, of the Sun-god (Shamesh).'"[57] Frothingham further noted "that, in an incantation, [Ningishzida] was referred to as the 'Herald of the Earth' and this had reference to his being the 'messenger of the Earth Mother.'"[58] Perhaps then we can say that it is Ningishzida who personifies the sub-lemniscatory *minera* as uroboric Mother Nature.

However, did we not see above that it is the *lemniscate* that interpreters of James Joyce have linked to the "eternal feminine" and to "earth goddesses"? We also found that Johnson related the *vesica piscis*, a form of the lemniscate, to the "Great Mother." Yet these lemniscatory associations, focusing as they do on the genitalia, emphasize the "life-giving and regenerative role" (Boyle) of the feminine, "the feminine principle of generation" (Blair). What I propose is that the *sub*-lemniscatory Earth Mother symbolized by the caduceus connects us to an even deeper expression of universal femininity, one that goes beyond primal sexuality and instinctive life to encompass the *fusion of life and death*. So, whereas the lemniscate is tied to the "feminine principle of generation," the

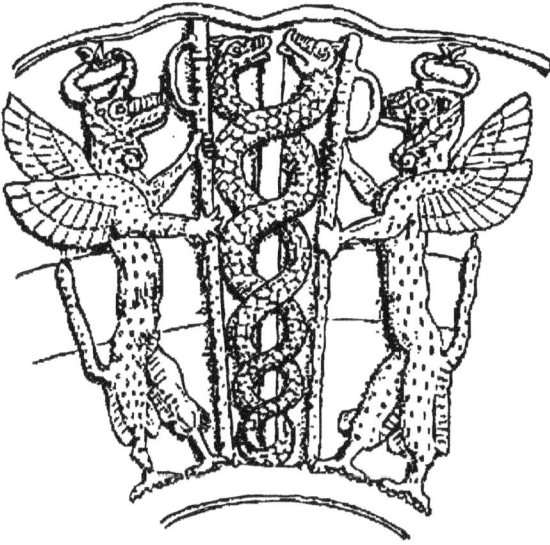

Figure 7.9 The goddess Ningishzida.

sub-lemniscate can be related to the most deeply chthonic principle of death and rebirth personified by uroboric Mother Earth.[D2]

But is there an anachronism operating here? For while the earliest evidence of the sub-lemniscatory caduceus apparently dates back only some 4,000–5,000 years to ancient Babylonia, the lemniscatory Venus figurines are considerably older, originating in the Upper Paleolithic era as long ago as 35,000 years. Given the sub-lemniscate's status as the most primordial member of our topo-dimensional family, should we not expect its uroboric manifestation as the intertwined serpents of the caduceus to predate any manifestations of the lemniscate?

To begin, let me note in general that "snake motifs are found everywhere in Upper Paleolithic European art" (Vakras), and the "coils, spirals, and meanders of the serpent appear in the very earliest art forms" (James).[59] Writer Linda Foubister even claims to offer an example in which serpent and Venus figurine are combined. Foubister concludes that "the Goddess herself may have taken the form of a serpent, or she may have been accompanied by a serpent. One of the earliest examples of the Serpent Goddess figure was a painting of a large fanged serpent in the cavern of La Baume-Latrone, France (about 26,000 BCE), with a body resembling the rounded curves of a woman."[60] The question for us, of course, concerns the specific configuration that the serpent takes, whether lemniscatory or sub-lemniscatory. Is there any evidence that the sub-lemniscatory caduceus dates back more than four or five thousand years? Retief and Cilliers say that "the origin of the double-snake emblem is shrouded in the mists of antiquity."[61] But perhaps the mist can be lifted to some degree.

In *The Time Falling Bodies Take to Light*, cultural historian William Irwin Thompson comments on a 17,000-year-old painting (fig. 7.10) found at Lascaux, the famous site of Paleolithic cave art.

Figure 7.10 *The Shaft of the Dead Man*. Upper Paleolithic cave painting discovered at Lascaux (adapted from photograph by Peter80, Wikipedia.org).

For our purposes, here are the key features of this enigmatic image: on the right is a bison whose intestines appear to be hanging down; on the left, a pole or staff with a bird on top; and in between, a man with a bird-like head who is seemingly lying on the ground. Thompson challenges the conventional interpretation of this picture as a hunting scene depicting an injured animal and slain hunter. Citing the claims of Joseph Campbell and Robin van Löben Sels, Thompson proposes that the bird-man is not a dead hunter but a shaman erotically aroused (note his erect penis) and in an ecstatic trance as he communes with the bison, who personifies a feminine deity: "If the man is a shaman, and the bison is an image of the Great Goddess, then the bison could be an epiphany of the Goddess coming to the shaman in the power vision."[62] Next Thompson observes that the staff mounted by the bird head "seems related to the fact that the man is shown with a bird's head. The staff with a bird on top, whether as totem pole or caduceus, is an ancient and universal symbol, and this painting from Lascaux may in fact be an expression of the source in Paleolithic religion from which all the later images [of the caduceus] derive."[63]

Thompson goes on to say:

> If van Löben Sels's intuitive hunch that the scene from the Shaft of the Dead Man is a shamanistic power vision and not a literal painting of a hunt is correct, then the ithyphallic condition of the man makes good sense. The erection does not have to do with fertility, although it is a sign of potency; the potency is expressed by the fact that the shaman in a deep trance state is in communication with the gods [or the nonhuman Earth Goddess]. If the bird on top of the staff is a sign of [shamanistic] transformation, then the bird head on the man indicates that he is one whose consciousness can fly into the sky . . . he is a shaman, an initiate who has won the favor of the divine feminine, the Shakti, the Great Goddess.[64]

This is all well and good, but there is a detail of figure 7.10 with crucial implications for the symbol of the caduceus that Thompson associates with the bird-pole: the windings spilling down from the right side of the bison. Thompson disagrees with the "straightforward and literal" interpretation of Lascaux specialist Annette Laming that these visceral coils are "entrails." He suggests instead that they constitute the "sign of the vulva" that signifies the essentially feminine nature of the Great Goddess bison with whom the bird-man ecstatically communes.[65] But to my eye, the viscera protruding from the bison do not have a specifically genital look as do features of the lemniscatory Venus figurines considered above, but have a serpentine uroboric appearance more characteristic of the *sub*-lemniscatory coilings of the caduceus. Indeed, intestines can themselves be symbolic of feminine deity and the snake-like entrails of the bison bring to mind the *alimentary uroboros* which, in signifying the *minera*, signifies Mother Earth. In fact, as I view the figure of the bison in the painting, my eye moves around its intestinal windings and up its hind side to its tail, which—if my imagination is not getting the better of me—seems to form a serpent's head! Assuming my eyes are not deceiving me and such a serpentine figure is actually present, perhaps the hiddenness of the figure, its blending into the body of the bison, speaks to the possibility that the image is inserting itself from the deepest recesses of the unconscious.[D3]

However, in the cave painting, we do not literally see the alimentary serpent wound around the bird-pole to complete the image of the caduceus. Taken by itself, the bird-pole is bare. What we must do to realize the caduceus is take the bird-pole and the serpentine coiling "as a couplet" (at one point in his text, Thompson says of the picture that "man and bison become a couplet").[66] By superimposing bird-pole and entrails in this way, we obtain the alimentary caduceus. It is interesting that, while Thompson relates the bird-headed staff to the caduceus and goes on to underscore the importance of the serpent in prehistoric religious imagery, he makes no mention of any serpentine images in the *Shaft of the Dead Man* itself. But this is precisely what is required for the caduceus to gain full expression in the painting.

Let me add a few comments on the extraordinary nature of *Shaft of the Dead Man*. Though the artist seems to have painted the elements of the scene as separate from one another (bison, man, bird-pole, intestines), under the influence of a weakly differentiated magical consciousness he or she may well have tended to experience them as merging and losing their distinctness. And yet coherent images do appear here, reasonable representations of objects in nature, depictions containing locally delineated parts more or less properly proportioned and separated from each other in orderly spatial relation. If magical consciousness was more governed by intuition than by image, if visual images were indistinct at best in the spatially incipient magical awareness that prevailed in Paleolithic times, how is it possible that a Paleolithic artwork dating back 17,000 years could feature such coherent images? It is true that what we modern observers see when we view ancient objects through our post-Renaissance lenses is affected by those lenses. Thus I pointed out above that our way of viewing the Venus figurines is likely quite different from how they were experienced originally. Yet even when we take into account this observer effect, the images painted on the Lascaux wall are so lucid that it seems improbable that the artist herself did not experience some of this clarity, though she may indeed also have had the tendency to experience separate images as merged. So I ask again how such clarity was possible for an artist living in an age when magical consciousness was dominant. The answer may lie in the above-mentioned hypothesis offered by Thompson, Campbell, van Löben Sels, and others: the painter of *Shaft of the Dead Man* was not a typical member of Paleolithic society but was a shaman capable of visions that surpassed ordinary awareness. Therefore, if the ordinary Paleolithic person functioned in a sleeplike magical state (Gebser) wherein clear-cut visual images were essentially lacking, we can conjecture that the shamanic artists of the time were *extra*ordinary individuals who could awaken to visual experiences transcending the world of magic.

Now, in dating the caduceus to the Lascaux cave of the Upper Paleolithic, have we gone far enough back? The existence of lemniscatory artifacts (the Venus figurines) from an earlier period of the Upper Paleolithic suggests that we need to go further still in order to confirm the association of the caduceus with the sub-lemniscatory structure of our most primordial uroboros, the alimentary. This movement back to origin takes us from "the feminine principle of generation" evidently so influential in the Upper Paleolithic to the very earliest expression of femininity, that emphasizing the principle of death and rebirth embodied in the Earth Mother.

In *The Great Cosmic Mother*, Monica Sjöö and Barbara Mor associate the Venus figurines with Cro-Magnon people (of the Upper Paleolithic) and say that the Neanderthal (of the Middle Paleolithic) had an even older goddess culture: "Much earlier [than the Cro-Magnon age], during the Neanderthaloid period (dating from at least 200,000 BC), evidence shows that great . . . power was attributed to the earth as Mother of Life and Death. . . . The dead were to reenter the earth (the tomb, the womb) to be reborn again."[67] Continuing:

> Entombments [in] . . . cave graves . . . were ways of returning bodies to the womb of Mother Earth, where they waited for rebirth. . . . Death is the powerful dramatic mystery equal to Birth—and both are overarched and contained by the Great Mother. . . . God *was female for at least the first 200,000 years of human life on earth.* . . . Wooden images of the Mother God were doubtless carved long before the stone Cro-Magnon Venuses, but wood does not survive.[68]

A little later, Sjöö and Mor bring in the important role of the serpent: "The snake was, first of all, a symbol of eternal life since each time it shed its skin it seemed reborn. . . . Gliding as it does in and out of holes and caverns in the earth, the serpent also symbolized the underground abode of the dead who wait for rebirth."[69] If we can conclude from this that the Earth Mother symbol of the Middle Paleolithic may have been accompanied by serpents or even that she herself assumed a serpentine form, might we go so far as to say that the serpents were *sub-lemniscatory,* that they took the form of the caduceus?

György Doczi appears to be speculating along these lines in *The Power of Limits.* Emphasizing the primacy of the double spiral, he intimates that it may have originally been related to coiling snakes. Next he associates the serpentine double spiral with mazes and with the intestine-like windings of the labyrinth: "These double spirals have been interpreted as symbols of death and rebirth, because as one follows the line coiling inward, one finds another line coming out in the opposite direction, suggesting both burial in the tomb and emergence from the womb."[70] In his next paragraph, Doczi seems to associate the double-spiral pattern at prehistoric gravesites with anthropologist Loren Eiseley's understanding of Neanderthal burial rites (though it is doubtful that Eiseley himself would have made this connection). Doczi then goes even further:

> Is it pure coincidence that on the molecular level the joint three-dimensional spiral pattern of the double helix—matching the double snakes of Hermes' magic wand [i.e., Mercury's caduceus]—was found a few years ago to be the true shape of the DNA molecule, which contains within its miniature coded pattern the master plan of the entire future development of living organisms? It was discovered even more recently that some of the most minute and important elements within living cell structures (such as red and white blood corpuscles) group themselves in double spiral patterns. The cores of these microtubules . . . are a faithful match to the double spirals of prehistoric tombs, the tattoos of the Maoris, and the Mother Earth patterns of the American Indians.[71]

For Doczi, then, the symbol of the caduceus could not be more fundamental, expressing as it does "the intertwining mystery of life and death."[72] But is there anything in the archaeological record to support the intuitive conjecture that the origin of the caduceus goes as far back as the Middle Paleolithic?

There is general agreement among archaeologists that the Upper Paleolithic brought a significant increase in symbol making. Anthropologists Roger Lewin and Robert Foley note that the development of art and symbol creation in the *Middle* Paleolithic is disputed, with some researchers insisting that little such activity took place. "Some evidence has been gathered to indicate the existence of image making earlier than the Upper Paleolithic, but it is very limited," say Lewin and Foley.[73] Their prime example of Middle Paleolithic symbolizing is "a fragment of bone marked with a zigzag motif, from the Bacho Kiro cave site in Bulgaria" (fig. 7.11).[74] (Lewin and Foley date this engraving to "somewhat earlier than 35,000 years ago" whereas Bahn and Vertut date it back 47,000 years.)[75]

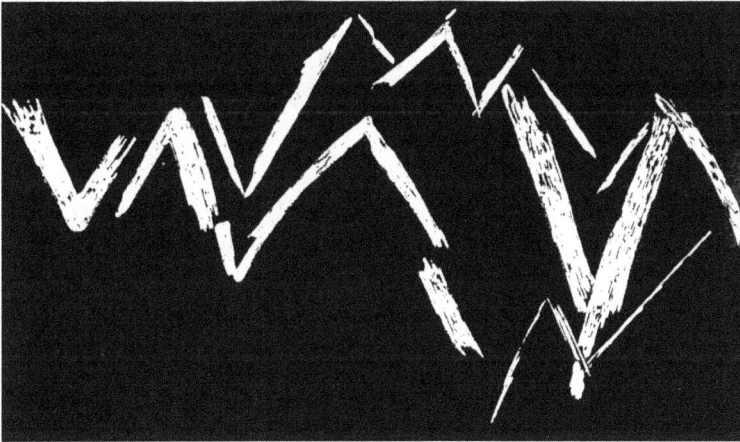

Figure 7.11 Middle Paleolithic zigzag motif found on fragment of bone, Bacho Kiro, Bulgaria (negative drawing).

In her book *The Language of the Goddess*, archaeologist Marija Gimbutas also cites the Bacho Kiro zigzag pattern, stating that the "zigzag is the earliest symbolic motif recorded; Neanderthals used this sign around 40,000 BC or earlier."[76] Gimbutas first relates the zigzag to water—"the image of water is zig-zag or serpentine"— then proceeds to associate it explicitly with the snake and Snake Goddess.[77] A number of other researchers identify the zigzag as a symbol of the serpent.[78] Then can we not say that the Middle Paleolithic figure carved on the Bacho Kiro bone might be symbolic of serpents? And if we imagine the jagged edges of the image smoothed out and the lines connected (fig. 7.12), would it be going too far to suggest the possibility that we might have here an original engraving of the coiling serpents of the caduceus?

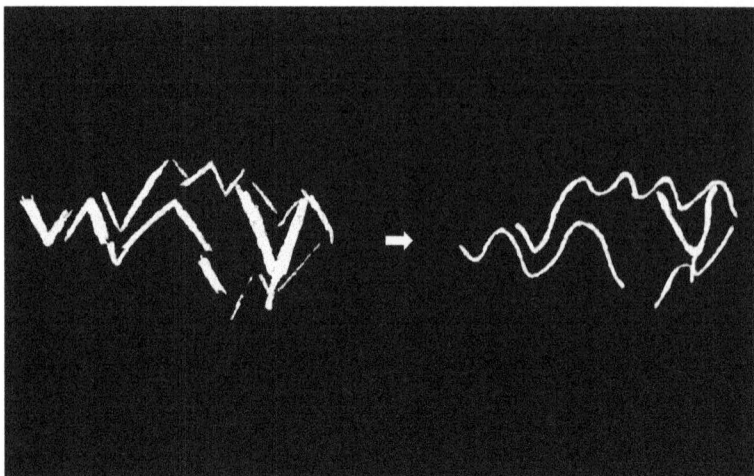

Figure 7.12 Smoothing of Bacho Kiro zigzag motif.

KUNDALINI DESCENDING: A SUMMARY OF
THE CONJUNCTIONS

We have just seen that William Irwin Thompson interpreted the bird-pole depicted in *Shaft of the Dead Man* as a precursor of the modern caduceus and the bird-man as a shaman in trance. Thompson alternatively described the bird-man's experience as an "awakening of *kundalini*."[79] He then related the caduceus to kundalini in a direct way by identifying the snakes on the staff of Mercury with the two snakes of Tantric yoga, of which kundalini is a part.[80] Joseph Campbell and many other scholars and commentators have confirmed this relationship.[81] I would now like to use the dynamics of the kundalini to reprise in summary form the topo-alchemical conjunctions set forth in this book.

The term *kundalini* is based on a Sanskrit word that means "spiraled" or "coiled," as in the coiling of a snake. In the tradition of kundalini yoga, a coiled serpent with its tail in its mouth is said to lie sleeping at the base of the spine. This uroboros is not merely a physical creature, of course, but is a life force that can animate the subtle body. When the kundalini serpent is awakened, it rises from its source in the perineum (the region between the anus and genitalia) and moves upward, successively activating concentrations of vitality known as "chakras." In Sanskrit, the word *chakra* means "wheel," and, as the serpent uncoils, energy is imparted to the wheel-like vortical centers that constitute the subtle body. This, in turn, is said to bring about heightened awareness and transformations of consciousness.

The best-known approach to kundalini yoga is the classical Hindu system positing seven chakras distributed vertically along the spine. Above the root chakra located at the spine's base, there are whirling centers of subtle energy found around the genitals, navel, heart, throat, forehead, and at the crown of the head. Note though that less well known alternative models of kundalini posit different numbers of chakras (different forms of Tantric Buddhism, for example, identify anywhere from three to ten), and varying accounts are also given of the relative importance of certain chakras, the specific functioning of the chakras, and even of the manner in which the kundalini energy flows. In my own account, I propose four main chakral centers (fig. 7.13), corresponding to the four body centers we have worked with: brain, heart, genitals, and intestines. Indeed, in the previous section, we saw that these four organ systems are aligned with our four manifestations of the uroboros.

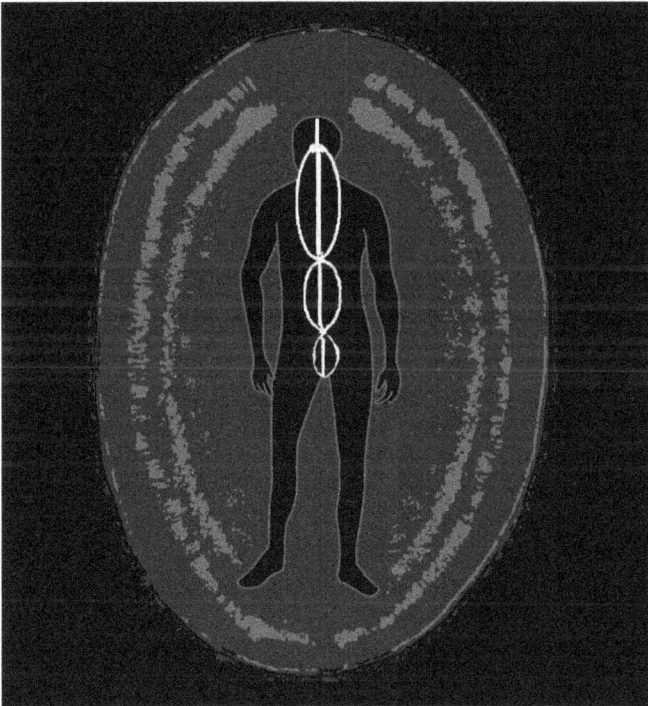

Figure 7.13 Flow of kundalini energy through four chakral centers, the chakras being located at the four points of intersection with the *sushumna* (the central column) (adapted from Pierjasi at nl.wikipedia).

Although classical approaches emphasize the upward movement of the kundalini energy, Jung suggests that, in the contemporary West, we must move *downward*. This is brought out in Jung's *Visions* seminar, in the course of discussing a vision in which a woman goes down a deep shaft to the bottom of a well and there encounters a coiled snake. Jung identifies the shaft as the *sushumna*,

> the canal through which the Kundalini rises. And here we have the remarkable fact that she is coming down. Here we see the tremendous difference between India and the West. You see, if she tried to go up the *sushumna*, it would be perfectly unnatural, a merely imaginary enterprise. The point is that she is already up above . . . she must come down. While the East is already below and has to establish a connection with the thing above, because clearness of consciousness does not exist with them, their consciousness is blurred. Therefore the great mistake which Western people make is imitating the Eastern yoga practices, for they serve a need which is not ours; it is the worst mistake for us to try to get higher and higher. What we should do is establish the *connection* between above and below.[82]

(I hasten to add that, while I agree with Jung's emphasis on the present need to move downward, his stark contrast between East and West seems simplistic to me. The flat claim that Eastern consciousness is "blurry" while Western consciousness is clear grossly overstates the case and fails to do justice to the complexity of the contemporary Eastern mindset. By the same token, we do not have to go back too far into *Western* history to find considerable blurriness [e.g., old alchemy].)

Later, again stressing that the challenge for the Western intellect is to move downward, Jung says: "That is in accord with all that I have found in practical analysis; in every case, without exception, it must be a descent because it is typical of the Western mind that it moves in a conscious world."[83] This "conscious world" is of course the world we have spoken of before, that in which the uroboros has been repressed in the interest of individuation. And what Jung, Neumann, Schwartz-Salant, and others have advised is that we must now make contact with the uroboros if individuation is to be brought to genuine fruition. This requires going *down*.

Over past millennia, the uroboric kundalini had indeed gone *up*, rising from its base where it had been sleeping loosely coiled. Awakening, tightening and intensifying, it had manifested fewer and fewer coils. This ascendance of the serpent reflects the alchemical *solutio* we encountered earlier—the rising of consciousness from its sleep-like condition of intricately interwoven polymorphousness to one of detached monomorphous abstraction—as mirrored in human culture's passage from polytheism to monotheism (see chapter 4), and in the transition from the child's entanglement in the multiple identities of his or her "imaginary playmates" to the sharply focused, unitary identity of the adult ego (see chapter 5). The movement up the *sushumna* also corresponds to the movement up our alchemically rendered topological series (fig. 7.14).

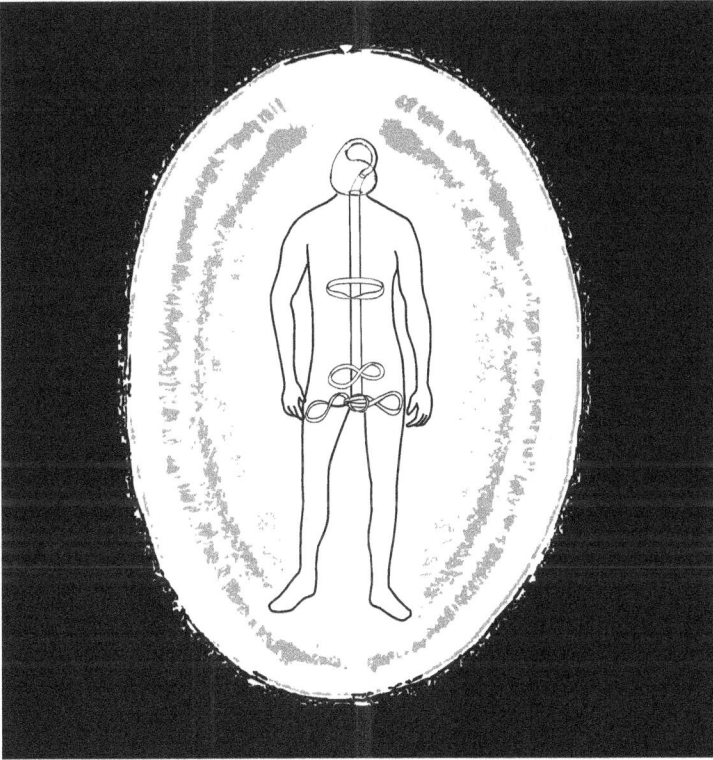

Figure 7.14 Schema of topological chakras in ascending order: sub-lemniscate, lemniscate, Moebius surface, and Klein bottle.

The ascending stages of development of the Klein bottle are reflected in figure 7.14. In the first stage, the embryonic Klein bottle bears the stamp of the sub-lemniscate. This incipient Kleinian structure is associated with the intestines and neural chassis, and is symbolized by the caduceus. Next, the Klein bottle takes the form of the doubly looped lemniscate. Here the primordial Klein bottle is related to the genitalia and reptilian brain, and is represented by the lemniscatory uroboros (fig. 7.4). The third stage casts the Klein bottle as the singly twisted Moebius. This ancient Kleinian form is linked to the heart and paleo-mammalian brain, and it finds expression in the old Egyptian uroboros (fig. 5.8). In the final stage of ascent, there is the modern Klein bottle, a structure associated with the mature human brain, embodied in the neo-mammalian cerebral cortex. What is now taking place is the *reversal* of the erstwhile Projective movement upward and forward that had constituted alchemy's *solutio*.

Entering the *coagulatio*, the gears are shifting to backward and downward. To be sure, this transition is no mere regression in which the rise of consciousness is simply undone. We saw in previous chapters that backward Proprioceptive action does not just cancel the forward movement of consciousness but operates against its grain, enabling us to experience the source of its still-occurring Projective thrust. Similarly, the going down that Jung called for in his discussion of kundalini must also be a going *up*—which is to say that it is not a regressive reversal of individuation, but a heightening of it that brings it to fulfillment. In this regard, anthropologist Herbert V. Guenther noted that, in attempting to understand the dynamic flow of kundalini energy in the *sushumna*, "it is only by grasping the dialectical interplay between 'top and bottom,' 'head and tail,' 'apex and base' within this scheme that one avoids the splitting-off of the one pole from the other."[84] Let us also take note of the dialectical interplay between particular chakras and the entire chakra system. This is especially evident in the case of the sub-lemniscatory chakra. For while this alimentary organ of the subtle body is uniquely symbolized by the caduceus, the overall layout of subtle organs—with its doubly serpentine twining around a central staff—is caducean as well!

We know the first step in the *coagulatio*: the "gears shift" and we move Proprioceptively backward into the brain, recognizing that—indeed, as a chakra system—the brain is more than just the finite physical object it has been Projected to be. Instead it is *psycho*physical or *sub*-objective, a Kleinian body of paradox twisting to infinity, a uroboric subtle organ. Such a realization constitutes the initial *coniunctio*, associated earlier with the *unio mentalis*. To advance to the second stage of conjunction, a certain "knot" in the Kleinian chakra needs to be undone.

The term *knot* is found in the Tibetan Buddhist approach to kundalini yoga wherein blockages or knots that have formed in the chakras as a result of one-sided attachments must be loosened by meditative practices that free the flow of energy. Thus art historians John C. Huntington and Dina Bangdel speak of Buddhist meditations that balance and center the passage of energy through the subtle body, "thereby loosening the knots of obfuscation, and awakening the chakras."[85] In the present account, it was during the process of Projection accompanying the initial rise of the kundalini serpent that the "knots" in the chakras had been formed: with each successive chakra activation, the previous chakra had become de-energized or repressed, tightly closed off as if tied up by a knot. Topologically, this is generally expressed by the upward movement in which the lower-dimensional members of our geometric series had originally become enclosed within the emergent higher-dimensional members—the sub-lemniscate within the lemniscate, the lemniscate within the Moebius, the Moebius within the Klein bottle—akin to the way *matryoshka* dolls are nested within each other. But, from what we have already said, it is clear that the pattern of repressive containment whereby "knots" are formed is more complex than that.

For one thing, it is not just the Moebius that is repressively enclosed within the Klein bottle during the stages of Projection, but the two lower-dimensional members of the bisection series as well. In terms of the stages of evolution of the 3 + 1-dimensional Klein bottle summarized above, each successive transition upward brings repressive containment within the Kleinian vessel of a lower-dimensional structure. In the second stage, the zero-dimensional sub-lemniscate is so concealed, in the third stage the 1 + 1-dimensional lemniscate, and in the fourth, the 2 + 1-dimensional Moebius. Each successive repression can be seen in light of the old alchemical tale of the "Spirit in the Bottle." With each there is a Kleinian "bottling up" of a different topo-dimensional manifestation of the shape-shifting "Spirit Mercurius," who is, after all, serpentine, as his staff attests. Moreover, if the evolution of the Klein bottle marks the "knot-creating" individuation of the human *cogito*, we should expect similar evolutionary processes in the nonhuman individuation of the Moebial *anima* and lemniscatory *vegeta*. Thus, the sub-lemniscate and lemniscate should not only have become repressively contained within the Klein bottle in the course of Projection but also, within the Moebius, and the sub-lemniscate would likewise have been contained within the lemniscate.

With the gears having now been switched to Proprioception and the *unio mentalis* carried out, the first topo-alchemical knot to be undone is the one in the Klein bottle that has constricted the lower-dimensional

Moebius chakra. The cutting of the Kleinian knot is enacted in the *bisection* of the bottle. But this cannot be an act of brute force like that carried out on gross matter, for—in freeing the hitherto confined Moebial uroboros, in liberating the dimension of feeling by opening the heart chakra (the *anahata*, in the Hindu kundalini system)—we are dealing with *subtle* matter. Let us say that the Klein bottle must be bisected in a manner similar to the division carried out by the August Master of the Center, a deity of the Shinto religion described by Nahum Stiskin in his book, *The Looking-Glass God:*

> Dwelling at the point of the splitting of the unitary energy of life, [the August Master] can be said to wield a divine sword which slices that energy into its two manifestations and thereby creates polarity. This deity, however, is the consummate swordsman who, although cutting into two, does so with such speed and precision that the fluid of life continues to flow between the resulting halves. They therefore remain continuous and intertwined.[86]

In the same way, when the Klein bottle is divided to yield the Moebius, the cut must be so "fast and precise," so subtle, that it is at once also *not* a cut and the bottle remains intact. As a consequence of such a paradoxical incision, the Proprioceptive descent into the world of the *anima* that constitutes the second *coniunctio*—far from entailing a loss of individuation in which the Kleinian *cogito* is simply left behind (as would happen with the ordinary, gross bisection of the Klein bottle)—brings *cogito* and *anima* into close synchrony. Like the Hindu deities Shiva and Parvati, these serpentine vortices of chakral energy twine sinuously together in an intimate cosmic dance.

The next knot in the Klein bottle to be subtly severed frees the lemniscatory uroboros and opens the chakra of chthonic sexuality (roughly corresponding to the Hindu *svadhisthana*). However, since the lemniscate is also repressively contained within the Moebial *anima*, the knot in the Moebial chakra must be undone as well. With this achieved in the process of animal individuation, vegetal sensuality is non-regressively unleashed to join the gyrations of *cogito* and *anima* in a cosmic dance of three that constitutes the third *coniunctio*.

Finally, with the cutting of the last chakral knot in the Klein bottle, the most primal chakra of all is unfettered: the sub-lemniscatory energy center associated with the intestinal windings of the alimentary uroboros, the original caduceus (fig. 7.11) that roots us firmly in the bowels of the earth (this "root chakra" is similar to the *muladhara* of Hindu tradition). Because the sub-lemniscate has also been "bottled up" within the Moebial *anima* and lemniscatory *vegeta*, the knots in these chakras too must be subtly severed. Upon accomplishing that, the ultimate *coniunctio* is realized, that which brings the *minera* into ecstatic transpermeation with the *cogito, anima,* and *vegeta* (fig. 7.15).

Figure 7.15 Topological adaptation of Emblem 17, from Michael Maier's *Atalanta Fugiens* (1617). In alchemy, four is the number of wholeness and archetypal conjunction. The four intersecting spheres of Maier's old alchemical work may be taken as symbolizing the conjunction of our four topo-alchemical energy spheres.

The paralleling of the kundalini process with the alchemical account of the *coniunctios* accords with the writing of Buddhist teacher and therapist Rob Preece. In *The Psychology of Buddhist Tantra,* Preece speaks of the union of the masculine and the feminine: "The *coniunctio,* as Jung called this union, is . . . central to the completion stage of Tantra, where the male and female aspects of the meditator are brought into union on an inner level."[87] Preece describes in detail the meditative blending of chakral energies in which the subtle energy of the crown, personified by the masculine deity (Heruka Chakrasamvara), is drawn down into the navel, whose subtle energy is related to the feminine deity, Vajrayogini. What results is an ecstatic conjunction of "god" and "goddess." Here "the body is seen as the alchemical vessel within which a powerful transformation takes place through the generation of bliss," "similar to the awakening of kundalini."[88] And Preece acknowledges that, when engaging in the embodied meditations necessary to achieve the awakening, "constricting knots" in the chakras must be dissolved.[89] In the *hieros gamos* that eventuates, the "duality that distinguishes mind from appearances, discriminating subjective mind and objective forms, disappears."[90] Of course, alchemy posits more than one such conjunction, and, in the present work, *several* have been described in portraying the transpermeations of our four topo-alchemical chakras.

In relating Tantric practice to alchemy, Preece articulates the central theme of *death and rebirth:* "The union of the masculine and feminine elements in the body leads to a kind of psychological death, a process of cleansing and transformation that totally and radically changes the inner reality of the individual. The meditator as he or she was dies, and a new relationship to reality dawns."[91] A key ritual of "Buddhist alchemy . . . called the 'inner offering'" involves a practice of visualizing one's death and rebirth, which is imagined to take place in a heated vessel called a skullcup.[92] "The visualization of the inner offering," says Preece, "is a metaphor for an alchemical process . . . that gradually unfolds in the body during meditation."[93] Moreover, "in Highest Yoga Tantra teachings, the process of death and reemergence is described in detail, with a clear understanding of how the energy body and the consciousness progress through the evolution of death, intermediate state, and rebirth. It is also recognized that we can experience a process in meditation that is virtually the same as actual death, except that consciousness does not leave the body. . . . These three phases of death, intermediate stage, and rebirth are experienced as a visualized simulation of what will take place . . . at the actual time of death."[94]

In preceding chapters of the present book, we have explored the alchemical ordeals of death and rebirth attendant to each of the *coniunctios:* death by decapitation (chapter 4), by drowning (chapter 5), and by fire (chapter 6). Now, in confronting the ultimate such ordeal, a final self-signification of the text is called for. But before turning to that, I conclude this kundalini-based summary of the *coniunctios* by noting that my four-decade journey to this place began with what I have come to understand as a spontaneous kundalini experience (the experience is also described in my commentary on the third dream in the chapter 2 dream journal).

On a night in 1968, I turned off the light beside my bed and drifted
into a hypnagogic vision of moving on a sled across a large field of
snow. Gliding soundlessly through the soft whiteness, a sense of serenity
enveloped me. The sled then began to rise into the air. As this happened,
I was seized by a feeling of exhilaration that quickly built into the sense
that every cell in my body was about to explode in unbearable ecstasy!
The painful bliss became so excruciating that, by a conscious effort of will,
I shut the experience down and went to sleep.

SELF-SIGNIFICATION OF THE ARCHAIC EMBRYO

To begin, I want to emphasize the distinction between a symbol like the caduceus that signifies something that goes beyond itself, on the one hand, and a *self*-signification, on the other. The caduceus can evocatively allude to archaic consciousness but, as a visual metaphor, it cannot embody that nonvisual awareness in a concretely self-referential way. As a consequence, it cannot distill the archaic right here in this text. Such a distillation is what we are presently seeking.

In each of the three previous chapters, we reached a point where we were faced with the task of doing more than just writing *about* the *coniunctios*. In order to make them a concrete reality, we had to distill them tangibly in the text itself via self-signification. We have now arrived at the same point with regard to the final conjunction. More is needed than just writing about mineral dimensionality and archaic consciousness, or even evoking them metaphorically through visual symbols. To seal the Kleinian vessel for the last time and liberate the *minera*, a new act of concrete self-signification is required.

As noted previously, the kind of self-signification that is called for must include the embodied self of him who signifies, namely, the author of this text. And while a new self-signification is indeed at hand, earlier enactments are still ongoing. In the mental self-signification or *unio mentalis*, I have been seeking to discard my cloak of anonymity and make myself present on the page (fig. 4.3). Moreover—though it flies in the face of my commonsense perception of myself—I continue my effort to grasp the truth that the egoic "I" seeming to define the center of "my" personhood is in fact a Projection of a *trans*personal Self. In fathoming this truth most concretely, the Projection of "Steven Rosen" is Proprioceptively withdrawn. The backward passage through the master signifier of the text—through the written "I"—into the neo-mammalian brain of humanity at large has been aided by meditation on the Kleinian signifier (fig. 4.2). This palpable retro-sensing behind the "I"/eyes that gives access to the neo-cortical activity of the Kleinian brain does not just bring to light a finite chunk of matter housed as mere object within the cranium of a particular organism. Instead it elucidates what Burrow called the "phyloörganism," the species-wide organism or communal body that in fact constitutes a uroborically non-finite subtle body. To the extent that the Proprioception in question has actually taken place, authorship of the text has passed from Steven Rosen's particular being to the generic Kleinian Self.

Coupled with the mental self-signification is the *mythic,* and the text is further coagulated by continuing to work with the hitherto repressed subtextual child, *Stevie* (fig. 5.5). As in chapter 5, the Proprioception of the oneiric child, aided by meditation on the ancient Egyptian precursor of the Klein bottle (fig. 5.8), seeks to follow a backward path into humankind's paleo-mammalian brain; in so doing, Kleinian authorship of the text is consolidated more tangibly. Also still in progress is the *magical* self-signification initiated in chapter 6. This is the process of self-reference that seeks to make present the even more primal, deeply subtextual being of instinct and raw sensuality—"Baby Stevie," and—through its rudimentary self-intuition, to Proprioceive infantile consciousness in the reptilian brain of generic humanity.

What of the new self-signification presently at hand? We learned in the second section of this chapter that the Self now to be signified differs in an important way from the mental, mythic, and magical Selves that have been Proprioceived in foregoing chapters. The *archaic* Self is not simply a positive being but participates in the selflessness and negativity of the *minera* in which it is enmeshed. It is true that the magical infant, for its part, possesses barely a trace of individuality. Yet, however feeble the infant's ego, it does constitute a positive order of being, whereas archaic individuality is but an embryonic potential embedded in the black hole of mineral nothingness. It is into this hole we must go for the self-signification that is called for.

Let me pave the way for entering the "realm of the dead" by recounting a few dreams from the many I have had along these lines over the years. As with all my dreams, the following surely contain much personal material that could lend itself to interpretation. In the present context however, the dreams in question are offered more as "tuning forks" intended to set up resonances with the underworld through their archetypal soundings and images.

Dream of April 26, 1988:

I'm going to be operated on. It's a life-threatening situation and I have great anxiety at the prospect of facing my death. I'm waiting my turn in the operating room for a dangerous procedure on my left eye. I'm just about overwhelmed by dread, but am fighting it.

On the operating table, a woman is curled up in a fetal position in absolute terror. She's screaming. Fighting with the doctors. She has to be held down.

I can vividly see the dying countenance of another patient. His face is ashen, thin, shaped like a triangle, with very light eyes. The fellow's teeth have spaces between them, evidently a sign of his terminal condition. This man has completely accepted his death. He speaks peacefully from a place where he seems dead already. Speaking with great sincerity, he says he knows that the ultimate truth involves the Moebius strip.

Dream of April 14, 2011:

I am dying and my existential fear is mounting. It feels claustrophobic. It feels like I am being pulled down into a hole from which I can never, ever emerge.

A doctor is attending me. I have sores on two of my fingers, my left forefinger and a finger on my right hand. If I'd had them tended to earlier there might have been a chance of saving me, but it's too late now. They've turned yellow and are gangrenous.

I am panicked. The time is coming, getting closer and closer. The doctor will give me a sedative . . . A sense of roundness, roundness . . . He'll give me a sedative and I'll drift off and never be able to return. I can't reconcile myself to that. . .

Existential fear. Clinging, clinging to my identity. I can't lose my identity forever, without ever coming back.

I am being sucked down. Irreversibly sucked into a hole, and I cannot accept it. . .

Since the *unus mundus* now to be entered is not only the "realm of the dead" but the "potential world of the first day of creation," dreams of death and *rebirth*, organic regeneration, are of special interest. One numinous dream of this sort begins on a topological note, proceeds to the theme of death, and concludes with the promise of embryonic life.

Dream of October 23, 1995:

> *I've solved the great metaphysical problem of existence. The solution is inscribed on a long, ornate, mysterious sheet of paper, subtle and richly intricate, possessing a faintly distinguishable Moebius-like quality. I want to store it in a safe place away from the ocean, whose waves could destroy it.*
>
> *The remainder of the dream centers on the theme of death. My sister is there and she is grieving, perhaps the loss of her husband. And she herself is dying. My mother is also dying or has already died, and I too am dying. There is the feeling of death all around, the inevitability of death.*
>
> *But then the question of regeneration is raised. There's a vivid image of pieces or segments of organic material. The living matter is cut into several sections. Are these segments vital enough to be able to grow together to generate a life process? The sections are smooth, white or beige, and have a vibrant glow to them.*

Another such dream is the one with which I opened this book. Let me reprise it.

Dream of September 20, 2005:

> *I'm traveling in the entourage of a well-known and charismatic philosopher. We are in a large, outdoor area, participating in a public conference attended by many. It's a momentous occasion and there is a feeling of sprawling openness and extension.*
>
> *At some point in the dream, the philosopher seems to be quite sick and people are attending to him. In fact he is dying and this is indicated by a huge, egg-shaped mass protruding from his abdomen. The massive ovoid distension points downward toward the earth and has the color of a cabbage. There are strands of purple predominating but other colors as well, some white, and a bit of green. And the protrusion seems "pregnant," about to burst open.*
>
> *I want to help him and feel I can. I ask him if he's ever had this experience before and he smiles and says he has. So the philosopher is in trouble and he appears to be dying, but there's a smile on his face. It has happened to him before and his smile tells me that all is well.*

I believe that dreams such as these help us get beyond simply *writing about* the final conjunction. They promote the self-signification of archaic consciousness here and now, in this text we are presently working with. It is a question of distilling the text, coagulating it to a density sufficient for the embryo's presence here. The written word is not nearly dense enough, of course. The master signifier of the archaic text clearly must be more concentrated than the written "I." Nor will an image of the self suffice as it did for the mythic signifier, nor a self-intuition, as in the magical case. The master signifier presently needed is one that operates from the selfless intuition of the mineral unconscious. For, as I noted earlier in this chapter, archaic cognition, being embryonic, cannot simply function on its own, not even intuitively. Only through the auspices of the maternal matrix in which it is embedded can the nascent archaic Self be intuitively signified. So, for this text to be dense enough, one must sink into the blackness of the *unus mundus*, the matrix in which the *minera* selflessly intuits the embryonic Self as a scintilla of light. To be sure, this is no signification of a Self that simply is positive. It is a reference made to embryonic life out of the selfless pit of death. Coagulating the text for the final *coniunctio* means descending into that pit, dropping down into the darkest depths of the unconscious, immersing oneself in the sphere of death that is at once the "potential world of the first day of creation." The descent must of course be conscious. It cannot be a mere regression to undifferentiated oblivion, but a regression *in the service of the text* that brings with it the adult's full Proprioceptive awareness.

Proprioception of the embryonic core bears some resemblance to the process of re-birthing described by the transpersonal psychiatrist Stanislav Grof in his essay, "Modern Consciousness Research and Human Survival."[95] Grof ascribes great significance to the "perinatal" period of development, that occurring around the time of birth. He claims that "the perinatal process transcends biology and has important psychological, philosophical, and spiritual dimensions. It would be an oversimplification to interpret it in a mechanistic and reductionistic fashion."[96] Grof associates the initial stage in the birthing process with an experience of "overwhelming fear. . . . The very onset of biological birth is experienced as imminent vital danger and threat of enormous proportions."[97] Grof calls for a *reliving* of this birth process. As a psychiatrist influenced by Freud, he understands that, when a person is in crisis, the difficulty can only be fully addressed by returning to its "psychological roots."[98] In his view, however, we are currently embroiled in a multitude of crises the dimensions of which are not only personal, but also "transpersonal," since the dilemmas besetting us extend beyond the individual sphere to directly encompass humanity at-large.[99] For Grof, addressing this "global crisis" means returning to the "perinatal level of the unconscious and the dynamics of the death-rebirth process [which] represent a repository of difficult emotions and sensations."[100] In effect then, Grof is urging a movement of awareness that takes us back down to our archaic origin in a manner that is transpersonal, as well as personal.

How is this to be achieved? According to Grof, "There exists a wide spectrum of ancient and Oriental spiritual practices that are specifically designed to facilitate access to the perinatal and transpersonal domains."[101] Among these are "various shamanistic procedures, aboriginal rites of passage and healing ceremonies, death-rebirth mysteries, and trance dancing in ecstatic religions."[102] It is even possible for such practices to bring cognizance of the moment of conception: "many people experience very concrete and realistic episodes which they identify as fetal and embryonal memories. It is not unusual under these circumstances to experience (on the level of cellular consciousness) full identification with the sperm and the ovum at the time of conception."[103] Grof goes on to emphasize the *phylogenetic* aspects of the rituals in question: "one can transcend the limits of the specifically human experience and identify with the consciousness of animals, plants, or even inorganic objects and processes."[104] From the outset he makes it clear that experiences like these cannot simply be dismissed as hallucinatory epiphenomena of the individual human brain. Although the "traditional model of the human psyche that dominates academic psychiatry is personalistic . . . observations of the last few decades have drastically changed our understanding of . . . the psyche."[105] This has resulted in a greater appreciation of the world beyond the person as an isolated unit of matter (i.e., a finite particular object-in-space). In a concluding remark, Grof comments: "All we can say is that somewhere in the process of confrontation with the perinatal level of the psyche, a strange qualitative Moebius-like [!] shift seems to occur in which deep self-exploration of the individual unconscious turns into a process of experiential adventures in the universe-at-large."[106]

Grof's mention of the possibility of identifying "with the consciousness of animals, plants, or even inorganic objects and processes" underscores the idea that *human* consciousness is not the only kind. The encounter with the "death-rebirth mysteries" in the *unus mundus* should indeed not only be undertaken by the human *cogito* but by its nonhuman counterparts as well. Previously I intimated that the *cogito*'s Proprioceptions are accompanied by the co-Proprioceptions of the *anima* and *vegeta*.

We saw in the last chapter that the third Kleinian Proprioception that brings us back to the magical human infant is linked to a second Moebial Proprioception returning the *anima* to a more primitive stage in its own development, a transformation whereby loving affiliation reverts to primal ferocity. I now propose that the fourth Kleinian Proprioception through which the *cogito* descends to its archaic roots is coupled with a third Moebial Proprioception wherein the *anima's* own most primitive condition is reexperienced—not as a condition of primal ferocity, but of primal fear. Furthermore, in the final conjunction, Kleinian and Moebial Proprioceptions are synchronized with a new Proprioception of the lemniscatory *vegeta* involving the sense of touch (see *Topologies of the Flesh* for more detail on forms of animal emotion and vegetable sensation, and their relation to phylogeny). Note that each Proprioception in the climactic *coniunctio* entails an embryonic self-signification mediated by the selfless intuition of the *minera*. Can we identify the organ correlates of these embryonic processes?

The intuition of the archaic *cogito* evidently would reach consummation in the most primitive manifestation of the brain, the neural chassis. Assuming that the primary organ centers of the *cogito's* nonhuman counterparts—the animal heart and the vegetal reproductive center—also undergo development, we can surmise that the intuition of the primeval *anima* would find its home in the oldest chamber of the heart, and that the intuition of the rudimentary *vegeta* would be grounded in the most primordial reproductive structure. No doubt more could be said about the development of the nonhuman organ systems, but, at this juncture, I will attempt no further analysis. We are nearing the finale of our alchemical odyssey, and, in the interest of coagulating the text as much as possible, I want to approach the final conjunction in less abstract, more immediate terms—those of existential experience.

Beyond the ordeals of death and rebirth accompanying the *coniunctios* of previous chapters—decapitation, drowning, immolation—there is the ultimate confrontation with death. If regression into the fiery sensuous realm of the infant involves "turning up the heat" (chapter 6), in the concluding ordeal—that entailing a return to the prenatal world—there is an exposure to *cold*. The crisis the ego now confronts does not merely concern its intellectual, emotional, or sensuous life, but the core of its very being. The primitive emotion triggered by this existential endangerment is cold fear. More tangibly still, one experiences a suffocating sense of closeness at the prospect of being entombed. With the icy fingers of death tightening their hold, there is no longer any possibility of maintaining even a modicum of egoic detachment. Now, in this densest and blackest of underworlds, tightly imprisoned within the cold bowels of the *minera*, one is really "up against it." Grof portrays this as an "experience of cosmic engulfment." Thus returning to origin, "the individual faces a situation that can best be described as no exit or hell. He or she feels stuck, encaged and trapped in a claustrophobic nightmarish world."[107] In the alchemical literature, this final ordeal is depicted by the burial of the king and queen following their *hieros gamos* (the holy wedding that symbolizes the penultimate *coniunctio*).

I have experienced this ordeal of burial in my own oneiric voyages. In a 2008 dream, I observed a scene that made me tremble: a man and a boy had been crammed into a steel cabinet by sinister forces. "They'll suffocate in there!" I thought in alarm. I sketched my impression of their terrible suffering in recording the dream (fig. 7.16).

Figure 7.16 Man and boy in steel box. Sketch of an image in a 2008 dream.

In a more recent dream, I am dealing with a fellow I don't trust. I feel he's trying to maneuver me into doing something that will have grave consequences for me, but I'm not sure what he really wants to do until the end. My recording of the dream ends with words of panic: *What he really wants to do is trap me in a box and bury me alive in that box! It's too late, I can't stop it anymore . . . It's too late! I'm trapped!. . . . Just horrible; horrible, horrible, horrible.*

What is the challenge confronting this fearful, anguished ego of mine as I face my demise? It is to realize that the suffocating steel cabinet confining the man and his child, the burial box enclosing me in claustrophobic terror, are containers of a *Kleinian* kind. This means that the box will be my tomb, and, at the same time, a womb, a delivery room. But the tomb must be sealed *hermetically*. There can be no avenue of escape, no way out for this ego that would maintain itself untransformed, with the Projection of its finite particular being left intact. Steven must resign himself to his total destruction *as* a finite center of identity. The loss must be felt. The grief must be processed. The fear must be faced. It is in carrying out these daunting tasks that the Kleinian vessel is sealed hermetically for the fourth time. With the Projection of Steven consciously drawing backward to its most primal source, Steven's finite body succumbs to the ravages of death and is transformed into a subtle body, the infinite body of archaic humanity-at-large, the body of stone comprising the Kleinian tomb (fig. 7.17).

Figure 7.17 Philosopher's Stone as Kleinian tomb. Topological adaptation of Michael Maier's "Squaring the Circle," from Emblem 21 of *Atalanta Fugiens* (1618).

"This is a tomb that has no body in it," says the inscription on the enigmatic tomb of Bologna. It is "a body that has no tomb round it. But body and tomb are the same." There is no body in the crypt because death has reduced that mortal vessel to dust. What remains is only the eternal stone of the crypt. Yet this uroboric sepulcher is in fact the regenerated body itself. It thus has "no tomb round it."

DREAM JOURNAL

D1

Dream of December 28, 2008:

Throughout this numinous dream, I am trying to tell a dream. But it isn't clear what the dream is, or who I want to tell it to.

The dreamscape has a static quality. There is a sense of formal patterning, as if each of the figures has been set in its proper place for a timeless ceremony. Here is a reproduction of the diagram I drew upon awakening:

The diagram shows me standing alongside a building (I'm the figure on the lower right). Some distance along the façade, are two men (sketched above). To my immediate left is a door, and, as it swings open, there is a woman standing at the threshold. She is facing me.

I wake up with an association to the two men: "yani or yoni" and "lingum" are the words I write down. I've heard these terms before but am not sure what they mean. My vague impression is that they have to do with male sex organs.

Follow-up research tells me that "yoni" is the Sanskrit word for vagina, and "lingam" is related to phallus. In Hindu religion, the terms have spiritual significance, being associated with the goddess Parvati and the god Siva, respectively. Theologian David Kinsley describes the yoni-lingam relationship as expressing the erotically divine union of opposites.[108]

I look again at my diagram. Recalling from my Wikipedia research that the yoni has been characterized as the "divine passage," I wonder whether the open doorway can be understood as the vulva. Perhaps then, the male figure facing the doorway could be taken as the lingam or phallus and the two male figures above possibly as testicles. The dream diagram could then be seen as mapping the primordial rite of sexual union.

D2

Dream of April 18, 2001:

My closest friends are visiting me. Something terrible then happens: My friends are all slaughtered by a group of wild animals (lions, tigers, etc.). The creatures are led by a mysterious black woman. She wears an African-style robe and holds a staff.

I am utterly grief-stricken and weep inconsolably. There's a feeling of profound loss: how can those so close to me all be destroyed in this way?

Reflection upon awakening: the black woman seems archetypal—the Black Goddess, an instinctive force of nature mercilessly destroying my domesticated relationships and supports, and leaving me groundless and grief-stricken in the process.

It seems to me now that, if the Goddess is at play, if it is not merely Death Mother but Great Mother Earth who presides, the death would at once contain seeds for rebirth.

D 3

Dream of February 8, 1986:

A very large snake becomes coiled around my neck, turned in a double loop. Having determined that the creature is not dangerous, I let it stay in this position. Someone nervously observing me asks whether the serpent is a "c," meaning cobra.

At some point, my mother appears across the street and sees me with the doubly coiled snake wound around my neck.

NOTES

Foreword

1. Jean Gebser, *The Ever-Present Origin* (Athens: Ohio University Press, 1985).

2. I'd like to acknowledge conversations with Lydia Salant on the mysterious nature of the Klein bottle, and also on Jean Gebser's work.

3. Nathan Schwartz-Salant, *The Black Nightgown* (Wilmette, IL: Chiron Publications, 2007).

4. Ioan Couliano, *Eros and Magic in the Renaissance* (Chicago: University of Chicago Press, 1987).

5. Gebser, *The Ever-Present Origin.*

Preface

1. James Hillman, *The Dream and the Underworld* (New York: Harper & Row, 1979), 131.

2. Steven M. Rosen, *Topologies of the Flesh* (Athens: Ohio University Press, 2006).

3. See C. G. Jung, *Alchemical Studies*, vol. 13, *The Collected Works of C. G. Jung*, trans. R. F. C. Hull (Princeton, NJ: Princeton University Press, 1967); C. G. Jung, *Psychology and Alchemy*, vol. 12, *The Collected Works of C. G. Jung*, trans. R. F. C. Hull (Princeton, NJ: Princeton University Press, 1968); C. G. Jung, *Mysterium Coniunctionis*, vol. 14, *The Collected Works of C. G. Jung*, trans. R. F. C. Hull (Princeton, NJ: Princeton University Press, 1970).

4. See Steven M. Rosen, *Science, Paradox, and the Moebius Principle* (Albany: State University of New York Press, 1994); Steven M. Rosen, *Dimensions of Apeiron* (Amsterdam: Editions Rodopi, 2004); Rosen, *Topologies of the Flesh*; Steven M. Rosen, *The Self-Evolving Cosmos* (London: World Scientific Publishing, 2008).

5. Maxine Sheets-Johnstone, *The Roots of Thinking* (Philadelphia: Temple University Press, 1990), 42.

6. Anita Hammer, "Mirroring and the Topology of Theatre," presented in Cross-Disciplinary Seminar, Department of Social Anthropology, Norwegian University of Science and Technology, Trondheim, Norway, February 15, 1999.

7. Jung, *Mysterium Coniunctionis*; Jung's description in fact makes explicit only *three* stages, but I demonstrate that one of these is actually best regarded as comprising a pair.

8. The cultural epochs noted here—mental-rational, mythic, magical, and archaic—have been elucidated by philosopher Jean Gebser. See Gebser's *The Ever-Present Origin* (Athens: Ohio University Press, 1985).

9. The word *abstract* is from the Latin *abstractus*, "dragged away, pp. of *abstrahere*, to draw from or separate" (*Webster's [Unabridged] Dictionary*). *Analysis* is a word of Greek origin that means a "dissolving, a resolution of whole into parts; *ana*, up, back, and *lysis*, a loosing, from *lyein*, to loose" (*Webster's [Unabridged] Dictionary*). Or we may say equivalently that "analysis" connotes "a breaking up" (*American College Dictionary*).

10. In Jung's *Collected Works*, these are *Alchemical Studies* (1967), *Psychology and Alchemy* (1968), and *Mysterium Coniunctionis* (1970).

Chapter 1

1. David Lavery, "The Eye of Longing," *Re-Vision* 6 (1983): 22–33; Jean Gebser, *The Ever-Present Origin* (Athens: Ohio University Press, 1985).

2. Gebser, *The Ever-Present Origin*, 19.

3. Walter Ong, *Interfaces of the Word* (Ithaca, NY: Cornell University Press, 1977).

4. Martin Heidegger, "Modern Science, Metaphysics, and Mathematics" (1962), in *Martin Heidegger: Basic Writings*, ed. David F. Krell (New York: Harper & Row, 1977), 247–82.

5. Owen Barfield, *Saving of Appearances* (Middletown, CT: Wesleyan University Press, 1988), 95.

6. Nathan Schwartz-Salant, *The Mystery of Human Relationship* (London: Routledge, 1998), 11.

7. Erich Neumann, *The Origins and History of Consciousness* (Princeton, NJ: Princeton University Press, 1954), 10–11.

8. Ibid., 11–12.

9. Ibid., 12.

10. Ibid., 109.

11. Julian Jaynes, *The Origin of Consciousness in the Breakdown of the Bicameral Mind* (Boston: Houghton Mifflin, 1976), 72.

12. W. T. Jones, *A History of Western Philosophy* (New York: Harcourt, Brace & World, 1952), 29.

13. Neumann, *The Origins and History of Consciousness*, 15, 14, and 13, respectively.

14. Ibid., 45.

15. Ibid., 115.

16. Ibid., 113.

17. Ibid., 45.

18. Ibid., 73–74.

19. Ibid., 87–88.

20. Ibid., 86.

21. Ibid., 89.

22. Ibid., 94.

23. Ibid., 121.

24. Ibid., 221.

25. Ernest Becker, *The Denial of Death* (New York: Free Press, 1973).

26. Neumann, *The Origins and History of Consciousness*, 27.

27. Friedrich Nietzsche, "On the Vision and the Riddle" (1878), in *The Portable Nietzsche*, ed. and trans. Walter Kaufmann (New York: Viking Press, 1968), 271.

28. I later discovered to my great relief that I did not have cancer after all!

Chapter 2

1. Erich Neumann, *The Origins and History of Consciousness* (Princeton, NJ: Princeton University Press, 1954), 36.

2. Steven M. Rosen, *Dimensions of Apeiron* (Amsterdam: Editions Rodopi, 2004).

3. David Bohm, "The Bohm/Rosen Correspondence," in *Science, Paradox, and the Moebius Principle*, ed. Steven M. Rosen (Albany: State University of New York Press, 1994), 232.

4. Eugene T. Gendlin, *Focusing* (New York: Bantam, 1978).

5. See Alfreda Galt, "Trigant Burrow and the Laboratory of the 'I,'" *The Humanistic Psychologist* 23 (1995): 19–39.

6. C. G. Jung, *Mysterium Coniunctionis*, vol. 14, *The Collected Works of C. G. Jung*, trans. R. F. C. Hull (Princeton, NJ: Princeton University Press, 1970), 534.

7. For the link between the *unus mundus* and Mercurius, see Marie-Louise von Franz, *Number and Time* (Evanston, IL: Northwestern University Press, 1974), 269; for the connection between Mercurius and the uroboros, see C. G. Jung, *Alchemical Studies*, vol. 13, *The Collected Works of C. G. Jung*, trans. R. F. C. Hull (Princeton, NJ: Princeton University Press, 1967); see also John P. Conger, *Jung and Reich: The Body as Shadow* (Berkeley, CA: North Atlantic Books, 1988), 151.

8. Citation retrieved June 15, 2011, from http://nysticorax.tripod.com/mercurius.html, a website providing information about alchemy.

9. Von Franz, *Number and Time*, 271.

10. Ibid., 269.

11. Ibid., 270.

12. Ibid.

13. Ibid., 279.

14. Ibid., 280.

15. Ibid.

16. Ibid., 279, 280.

17. Ibid., 280.

18. Ibid.

19. Ibid., 280–81.

20. See Steven M. Rosen, *Topologies of the Flesh* (Athens: Ohio University Press, 2006), chapter 4.

21. Von Franz, *Number and Time*, 280.

22. Ibid., 271.

23. Neumann, *The Origins and History of Consciousness*, 88.

24. Jean Gebser, *The Ever-Present Origin* (Athens: Ohio University Press, 1985), 149.

25. Neumann, *The Origins and History of Consciousness*, 11.

26. C. G. Jung, *Psychology and Alchemy*, vol. 12, *The Collected Works of C. G. Jung*, trans. R. F. C. Hull (Princeton, NJ: Princeton University Press, 1968), 242–43.

27. Ibid., 277–78.

28. Ibid., 278–79.

29. Barbara Obrist, "Visualization in Medieval Alchemy," *HYLE—International Journal for Philosophy of Chemistry* 9, no. 2 (2003): 131–70.

30. Gebser, *The Ever-Present Origin*, 50.

31. Ibid., 48. Note that Gebser actually dates the magical structure of consciousness thousands of years back into antiquity. What I am suggesting is that the same pattern of thinking is likewise evidenced in medieval alchemy. In chapter 6, I will elaborate further on the magical structure as described by Gebser.

32. This section, "The Hermetic Vessel," derives from Steven M. Rosen, "Pouring Old Wine into a New Bottle," in *The Interactive Field in Analysis*, ed. Murray Stein (Wilmette, IL: Chiron, 1995), 121–41.

33. Jung, *Psychology and Alchemy*, 236.

34. For the reference to the *"vas rotundum,"* see Jung, *Mysterium Coniunctionis*, 279; Jung notes that the "roundness" must be "simple and perfect" in *Psychology and Alchemy*, 88.

35. Jung, *Psychology and Alchemy*, 167.

36. Ibid., 236, n.15

37. For the reference to the "dragon biting its own tail," see Jung, *Psychology and Alchemy*, 293; for the appearance of the uroboros in images of the Hermetic vessel, see ibid., 290; see also, Ralph Metzner, *Maps of Consciousness* (New York: Macmillan, 1971), 96; and John Read, *Prelude to Chemistry* (Cambridge, MA: MIT Press, 1966), plate 35.

38. Read, *Prelude to Chemistry*, 149.

39. Jung, *Alchemical Studies*, 148.

40. Ibid., 87.

41. Jung, *Mysterium Coniunctionis*, 81.

42. Ibid., 57.

43. Ibid., 63.

44. For the reference to the "vessel of diaphanous glass," see ibid., 202; reference to the glass vessel with "eyes" is found in Jung, *Alchemical Studies*, 86.

45. Jung, *Alchemical Studies*, 193–98.

46. Ibid., 193.

47. Ibid., 197.

48. Ibid., 202–3.

49. Jung, *Mysterium Coniunctionis*, 465.

50. Ibid., 474–75.

51. Ibid., 471.

52. Ibid., 316.

53. Gebser, *The Ever-Present Origin*, 14.

54. Ibid.

55. Ibid., 14–15.

56. Ibid., 15, 16.

57. See Rosen, *Dimensions of Apeiron*, for more detail.

58. For my in-depth examination of modern physics, see Steven M. Rosen, *The Self-Evolving Cosmos* (London: World Scientific Publishing, 2008).

59. Paul C. Vitz and Arnold B. Glimcher, *Modern Art and Modern Science* (New York: Praeger Books, 1984), 118.

60. John Berger and Jean Mohr, *Another Way of Telling* (New York: Pantheon Books, 1982), 86.

61. Benoit Mandelbrot, *Fractals* (San Francisco: Freeman, 1977), 13.

62. Historian of science Gerald Holton provides insight into the relationship between Kierkegaard and Bohr. See Gerald Holton, *Thematic Origins of Scientific Thought* (Cambridge, MA: Harvard University Press, 1988).

63. Søren Kierkegaard, "Truth Is Subjectivity" (1846), in *Existentialism from Dostoevsky to Sartre*, ed. Walter Kaufmann (New York: New American Library, 1975), 110–20.

64. Rosen, *Dimensions of Apeiron*.

65. Rosen, *The Self-Evolving Cosmos*.

66. Nathan Schwartz-Salant, *The Black Nightgown* (Wilmette, IL: Chiron, 2007).

67. For alchemy's personal influence on Jung, see C. G. Jung, *Memories, Dreams, Reflections* (New York: Vintage, 1989), chapter 7; the volumes Jung dedicated to alchemy include *Alchemical Studies, Psychology and Alchemy*, and *Mysterium Coniunctionis*.

68. Aniela Jaffé, *From the Life and Work of C. G. Jung* (Einsiedeln, Switzerland: Daimon, 1989), 72.

69. Jung, *Psychology and Alchemy*, 279.

70. Nathan Schwartz-Salant, *The Mystery of Human Relationship* (London: Routledge, 1998), 10.

71. Ibid., 14.

72. Ibid.

73. Ibid., 18.

74. Ibid., 5.

75. Marie-Louise von Franz, "Psyche and Matter in Alchemy and Modern Science," *Quadrant* 8 (1975): 42.

76. Schwartz-Salant, *The Mystery of Human Relationship*, 11.

77. Von Franz, *Number and Time*, 279.

78. Jung, *Mysterium Coniunctionis*, 535.

79. Ibid., 539.

80. Ibid.

81. For reference to the "first day of creation," see ibid., 537; for reference to the "*prima materia*," see ibid., 534; the correlation with the lapis can be found in Jung, *Alchemical Studies*, 235.

82. Von Franz, *Number and Time*, 271.

83. Marie-Louise von Franz, *On Dreams and Death* (Boston: Shambhala, 1987), 148.

Chapter 3

1. C. G. Jung, *Mysterium Coniunctionis*, vol. 14, *The Collected Works of C. G. Jung*, trans. R. F. C. Hull (Princeton, NJ: Princeton University Press, 1970), 81.

2. Nathan Schwartz-Salant, *The Mystery of Human Relationship* (London: Routledge, 1998), 13.

3. I have worked extensively with the cube in other forums; see Steven M. Rosen, *Science, Paradox, and the Moebius Principle* (Albany: State University of New York Press, 1994); Steven M. Rosen, *Dimensions of Apeiron* (Amsterdam: Editions

Rodopi, 2004); Steven M. Rosen, *Topologies of the Flesh* (Athens: Ohio University Press, 2006); Steven M. Rosen, *The Self-Evolving Cosmos* (London: World Scientific Publishing, 2008).

4. Schwartz-Salant, *The Mystery of Human Relationship*, 5.

5. I have explored this paradoxical structure for over four decades. See, for example, Steven M. Rosen, "A Plea for the Possibility of Visualizing Existence," *Scientia* 108, nos. 9–12 (1973): 789–802; Steven M. Rosen, "A Neo-Intuitive Proposal for Kaluza-Klein Unification," *Foundations of Physics* 18, no. 11 (1988): 1093–1139; Rosen, *Science, Paradox, and the Moebius Principle*; Rosen, *Dimensions of Apeiron*; Rosen, *Topologies of the Flesh*; and Rosen, *The Self-Evolving Cosmos*.

6. Rudolph Rucker, *Geometry, Relativity, and the Fourth Dimension* (New York: Dover, 1977).

7. See Edwin Abbott, *Flatland: A Romance of Many Dimensions* (1884; repr., New York: Barnes & Noble, 1983).

8. C. G. Jung, *Psychology and Alchemy*, vol. 12, *The Collected Works of C. G. Jung*, trans. R. F. C. Hull (Princeton, NJ: Princeton University Press, 1968), 128, n. 44.

9. C. G. Jung, *Alchemical Studies*, vol. 13, *The Collected Works of C. G. Jung*, trans. R. F. C. Hull (Princeton, NJ: Princeton University Press, 1967), 148.

10. Ibid., 197.

11. Jung, *Mysterium Coniunctionis*, 474.

Chapter 4

1. C. G. Jung, *Mysterium Coniunctionis*, vol. 14, *The Collected Works of C. G. Jung*, trans. R. F. C. Hull (Princeton, NJ: Princeton University Press, 1970), 471.

2. Ibid.

3. Ibid., 472.

4. Ibid.

5. C. G. Jung, *Psychological Types*, vol. 6, *The Collected Works of C. G. Jung*, trans. R. F. C. Hull (Princeton, NJ: Princeton University Press, 1971).

6. Ibid., 454.

7. Ibid.; the four citations from Jung are found on 538, 518, 539, and 519, respectively.

8. Jung, *Mysterium Coniunctionis*, 534.

9. Marie-Louise von Franz, *Number and Time* (Evanston, IL: Northwestern University Press, 1974), 271.

10. Jung, *Mysterium Coniunctionis*, 539.

11. For reference to "universal ego," see von Franz, *Number and Time*, 280.

12. Deleuze and Guattari argue similarly that, in modernism, "a new type of unity triumphs in the subject . . . a higher unity . . . in an always supplementary dimension to that of its object." See Gilles Deleuze and Felix Guattari, *A Thousand Plateaus: Capitalism and Schizophrenia* (Minneapolis: University of Minnesota Press, 1987), 6.

13. Jacques Lacan, "Of Structure as an Inmixing of an Otherness Prerequisite to Any Subject Whatever" (1966), in *The Languages of Criticism and the Sciences of Man: The Structuralist Controversy*, ed. Richard Macksey and Eugenio Donato (Baltimore: Johns Hopkins University Press, 1970), 194.

14. Gayatri C. Spivak, "Translator's Preface," in Jacques Derrida, *Of Grammatology* (Baltimore: Johns Hopkins University Press, 1976), xix.

15. Ibid.

16. Eugene T. Gendlin, "Words Can Say How They Work," in *Proceedings, Heidegger Conference*, ed. Robert P. Crease (Stony Brook: Department of Philosophy, State University of New York at Stony Brook, 1993), 34.

17. Ibid.

18. Ibid., 29.

19. Eugene T. Gendlin, "Thinking Beyond Patterns: Body, Language, and Situations," in *The Presence of Feeling in Thought*, ed. Bernard den Ouden and Marcia Moen (New York: Peter Lang, 1991), 116–17.

20. Charles S. Peirce, *Collected Papers*, vol. 2, ed. Charles Hartshorne and Paul Weiss (Cambridge, MA: Harvard University Press, 1933), 230; Paul Ryan, *Video Mind/Earth Mind* (New York: Peter Lang, 1993).

21. Jung, *Mysterium Coniunctionis*, 498.

22. Ibid., 497.

23. Ibid., 499.

24. C. G. Jung, *Psychology and Alchemy*, vol. 12, *The Collected Works of C. G. Jung*, trans. R. F. C. Hull (Princeton, NJ: Princeton University Press, 1968), 128, n. 44.

25. Ibid., 277–78.

26. Ibid., 279.

27. Maurice Merleau-Ponty, *The Visible and the Invisible*, trans. Alphonso Lingis (Evanston, IL: Northwestern University Press, 1968), 145.

28. The term *identity operator* derives from mathematics, where it denotes a number that leaves unchanged the number on which it operates. For multiplication, this indispensable grounding element is the number 1. The iconic similarity between the mathematical "1" and the English "I" is noteworthy—especially when we take into account the distinctively *phallic* character of these two prime operators suggestive of patriarchal projection!

29. Trigant Burrow, *Science and Man's Behavior* (New York: Philosophical Library, 1953), 526.

30. Ibid., 249–54.

31. Alfreda Galt, "Trigant Burrow and the Laboratory of the 'I,'" *The Humanistic Psychologist* 23 (1995): 31.

32. Burrow refers to the "solidarity of the species" in *Science and Man's Behavior*, 71.

33. Ibid., 526.

34. Ibid., 95.

35. Drew Leder, *The Absent Body* (Chicago: University of Chicago Press, 1990), 117.

36. This term is defined as the "species man regarded as an organismic whole in which the element or individual is a phylically integrated unit"; Burrow, *Science and Man's Behavior*, 531.

37. Diego Rapoport, "Surmounting the Cartesian Cut through Philosophy, Physics, Logic, Cybernetics, and Geometry," *Foundations of Physics* 41(2011): 45.

38. M. C. Escher, *The Graphic Work of M. C. Escher* (New York: Ballantine, 1971), 16 (emphasis added).

39. Leder, *The Absent Body*, 17.

40. Ibid., 13.

41. Reference to the *"I think"* is found in Merleau-Ponty, *The Visible and the Invisible*, 145.

Chapter 5

1. C. G. Jung, *Mysterium Coniunctionis*, vol. 14, *The Collected Works of C. G. Jung*, trans. R. F. C. Hull (Princeton, NJ: Princeton University Press, 1970), 474.

2. C. G. Jung, *Aion*, vol. 9ii, *The Collected Works of C. G. Jung*, trans. R. F. C. Hull (Princeton, NJ: Princeton University Press, 1959), 253.

3. Maurice Merleau-Ponty, *Phenomenology of Perception*, trans. Colin Smith (London: Routledge and Kegan Paul, 1962), 410. Note also philosopher Herbert Spiegelberg's comment that "Merleau-Ponty finally characterizes 'time as the subject and the subject as time.' By this he means that the subject is not simply *in* time, for it assumes and lives time and is involved in time: it is permeated with time"; Herbert Spiegelberg, *The Phenomenological Movement* (The Hague: Martinus Nijhoff, 1982), 567. Spiegelberg additionally portrays Merleau-Ponty's analysis of time and subjectivity as "an attempt to combine Husserl's phenomenology of time with that of Heidegger" (ibid., 566–67).

4. Jean Gebser, *The Ever-Present Origin* (Athens: Ohio University Press, 1985), 66.

5. Ibid.

6. Ibid., 165–73.

7. Ibid., 61–73 and 165–73.

8. Jung, *Mysterium Coniunctionis*, 132.

9. Ibid., 140, 143.

10. Ibid.

11. For the relation of astrality to the moon, see Rudolf Steiner, *An Outline of Occult Science* (Spring Valley, NY: Anthroposophic Press, 1972); the relation of astrality to water is discussed in Ralph Metzner, *Maps of Consciousness* (New York: Macmillan, 1971), 89.

12. C. G. Jung, *Alchemical Studies*, vol. 13, *The Collected Works of C. G. Jung*, trans. R. F. C. Hull (Princeton, NJ: Princeton University Press, 1967), 257.

13. Jung, *Mysterium Coniunctionis*, 472.

14. Ibid., 143–44.

15. Gebser, *The Ever-Present Origin*, 271. Note, however, that Gebser is not entirely consistent in associating emotion with the mythic structure; see Steven M. Rosen, *Topologies of the Flesh* (Athens: Ohio University Press, 2006), 167–68.

16. Antony Flew, *A Dictionary of Philosophy* (New York: St. Martin's Press, 1979), 14.

17. Ibid., 377.

18. C. G. Jung, *Alchemical Studies*, vol. 13, *The Collected Works of C. G. Jung*, trans. R. F. C. Hull (Princeton, NJ: Princeton University Press, 1967), 307.

19. Erich Neumann, *The Origins and History of Consciousness* (Princeton, NJ: Princeton University Press, 1954), 329.

20. Ibid., 331.

21. Michael Washburn, *The Ego and the Dynamic Ground* (Albany: State University of New York Press, 1988), 45, 47, 48.

22. Neumann, *The Origins and History of Consciousness*, 332.

23. Ibid., 331.

24. Ibid., 332.

25. Ibid.

26. Ibid.; reference to the "medullary region" is found on 331, "heart-soul" on 237.

27. Ibid., 414–15.

28. Ibid., 338.

29. Ibid., 144–45.

30. Ibid., 145.

31. Ibid., 337.

32. Ibid., 338.

33. Morris Berman, *Coming to Our Senses* (New York: Bantam, 1989), 68.

34. Ibid., 66.

35. Ibid., 69.

36. Ibid., 70.

37. Ibid., 71.

38. Ibid., 73.

39. Ibid., 85.

40. Ibid., 97.

41. Ibid., 91.

42. Ibid., 69.

43. Jules Lachelier, "Psychology and Metaphysics," in *The Search for Being*, ed. and trans. Jean T. Wilde and William Kimmel (New York: Noonday Press, 1962), 155 (emphasis added).

44. Ibid., 168, 172.

45. Ibid., 168.

46. Ibid., 169.

47. P. D. Ouspensky, *Tertium Organum* (New York: Vintage, 1970), 79–80.

48. Ibid., 89.

49. Ibid.

50. Jung, *Mysterium Coniunctionis*, 476.

51. The reference to "animal soul" is found in Flew, *A Dictionary of Philosophy*, 14; "transpersonal, spiritual being" comes from Neumann, *The Origins and History of Consciousness*, 145.

52. Neumann, *The Origins and History of Consciousness*, 12.

53. David Lavery, "The Eye of Longing," *Re-Vision* 6 (1983): 22–33.

54. Ibid., 27–28.

55. Owen Barfield, *Saving of Appearances* (Middletown, CT: Wesleyan University Press, 1988), 24.

56. Gebser, *The Ever-Present Origin*, 72.

57. Ibid., 146.

58. Walter Ong, *Interfaces of the Word* (Ithaca, NY: Cornell University Press, 1977).

59. Ibid., 10.

60. Ibid., 18.

61. Ibid., 20–21.

62. Gebser, *The Ever-Present Origin*, 252.

63. Mary Lynn Kittelson, "The Acoustic Vessel," in *The Interactive Field in Analysis*, ed. Murray Stein (Wilmette, IL: Chiron, 1995), 89.

64. Ibid., 90–91.

65. Ibid., 96.

66. Ibid., 103.

67. Ibid.; reference to "creation myths" is found on 91; the "myth of Echo" is discussed on 101.

68. Ibid., 101–2.

69. Ibid., 102.

70. A while back, I was rummaging through some old belongings and happened upon a book of mine from childhood. It was a colorful, richly illustrated volume featuring a large centerfold map of the United States showing logging in the Northwest, cotton farming in the South, and so on. I don't believe the book had even entered my mind since I was six or seven, and suddenly encountering its centerfold image in this way seemed to bring back my original experience of it. I could see vividly the logs splashing down the whitewater river, could feel the fluffy texture of the cotton in the field, could sense the warm sun beating down on me in the Arizona desert. To Stevie, this picture was not just an object on a page that he viewed with detachment, but a world in which he was immersed.

71. Paul D. MacLean, "Alternative Neural Pathways to Violence," in *Alternatives to Violence*, ed. L. Ng (Alexandria, VA.: Time-Life, 1968), 22–34. Although details of MacLean's theory have been challenged in some scientific circles, I believe the concept remains valid in its general outlines, and it will prove useful for the purposes of this presentation. As Baars and Gage put it: "While some aspects of the triune brain theory have been controversial, nevertheless, it remains a helpful way to think about different 'layers' of the mammalian brain"; Bernard J. Baars and Nicole M. Gage, *Cognition, Brain, and Consciousness: Introduction to Cognitive Neuroscience* (Burlington, MA: Academic Press, 2010), 422.

72. See Allen R. Braun et al., "Regional Cerebral Blood Flow Throughout the Sleep-Wake Cycle," *Brain* 120 (1997): 1173–1197; Allen R. Braun et al., "Dissociated Pattern of Activity in Visual Cortices and Their Projections During Human Rapid Eye Movement Sleep," *Science* 279 (1998): 91–95; Pierre Maquet et al., "Functional Neuroanatomy of Human Rapid-Eye-Movement Sleep and Dreaming," *Nature* 383 (1996): 163–66.

73. As regards light, in *Dimensions of Apeiron* I explored its special role in the human world: light is no mere object that is seen but is that *by* which we see. In discussing the implications of the famous Michelson-Morley experiment on light that gave impetus to Einstein's theory of relativity, I demonstrated that the phenomenon of light—instead of lending itself to being treated as an object open to the scrutiny of a subject that merely stands apart from it—must be understood as entailing the *inseparable blending of subject and object*, of seer and seen. For his part, phenomenological philosopher Martin Heidegger adumbrated an intimate linkage among lighting, thinking, and Being: to think Being is to think light; Martin Heidegger, "The End of Philosophy and the Task of Thinking" (1964), in *Martin Heidegger: Basic Writings*, ed. David F. Krell (New York: Harper & Row, 1977), 370–92.

74. The reference to what "primitive man shares with the animal" is from Neumann, *The Origins and History of Consciousness*, 329; mention of the "resonant" and "ear-minded" character of the world of feeling is found in Kittelson, "The Acoustic Vessel," 90.

75. Julian Jaynes, *The Origin of Consciousness in the Breakdown of the Bicameral Mind* (Boston: Houghton Mifflin, 1976), 72.

76. Ibid.

77. Ibid., 73–75.

78. Ibid., 94–95.

79. Ibid., 95.

80. Ibid., 96.

81. Ibid., 396.

82. Mircea Eliade, *Myths, Dreams, and Mysteries* (New York: Harper & Row, 1960), 63.

83. Ibid., 177.

84. Ibid., 59.

85. Ibid., 60.

86. Ibid., 62.

87. Ibid., 61–62.

88. David Applebaum, "On Turning a Zen Ear," *Philosophy East and West* 33 (1983): 117.

89. Ibid.

90. The reference to "activity, unfolding, doing" appears in ibid., 118.

91. Ibid., 119 (emphasis added).

92. Ibid., 120.

93. Ibid., 121.

94. Eliade, *Myths, Dreams, and Mysteries*, 61.

95. Ibid., 102.

96. Neumann, *The Origins and History of Consciousness*, 145.

97. See Susan Braine, *Drumbeat/Heartbeat: A Celebration of the Powwow* (Minneapolis: Lerner Publications, 1995).

98. Lisa Sloan, "Shamanic Initiation: Map of the Soul" (doctoral dissertation, Pacifica Graduate Institute, 1999), 83.

99. Note that both Eliade and Jung acknowledged the parallels between shamanism and alchemy. See Eliade, *Myths, Dreams, and Mysteries*; Mircea Eliade, *The Forge and the Crucible* (New York: Harper & Row, 1962); Mircea Eliade, *Shamanism* (Princeton, NJ: Princeton University Press, 1964); and Jung, *Alchemical Studies*.

100. Eliade, *Shamanism*, 76.

101. Ibid., 236.

102. Among the most grueling accounts of this kind is that given in Jung, *Alchemical Studies*, 57–105.

Chapter 6

1. And the Moebius vessel is sealed a second time, but, as noted in the postscript to the last chapter, unpacking this detail of animal individuation is beyond the scope of the present work. For a fuller account of nonhuman individuation, see Steven M. Rosen, *Topologies of the Flesh* (Athens: Ohio University Press, 2006).

2. Jean Gebser, *The Ever-Present Origin* (Athens: Ohio University Press, 1985), 50.

3. Ibid., 48.

4. Ibid., 46.

5. Ibid., 47.

6. Ibid., 48.

7. Ibid.

8. Ibid., 46.

9. Ibid.

10. Ibid., 50.

11. Ibid., 149.

12. Ibid., 162ff.

13. Ibid., 62, 66.

14. David Abram, *The Spell of the Sensuous* (New York: Vintage, 1996), 188–89.

15. Ibid., 191–92.

16. Antony Flew, *A Dictionary of Philosophy* (New York: St. Martin's Press, 1979), 377.

17. Mircea Eliade, *Myths, Dreams, and Mysteries* (New York: Harper & Row, 1960), 59.

18. Mircea Eliade, *Shamanism* (Princeton, NJ: Princeton University Press, 1964), 91–92; see also Erich Neumann, *The Origins and History of Consciousness* (Princeton, NJ: Princeton University Press, 1954), 145.

19. Eliade, *Shamanism*, 118, n. 17.

20. Ibid., 474–77.

21. Ibid., 475.

22. Ibid.

23. Ibid., 476–77.

24. Eliade, *Myths, Dreams, and Mysteries*, 65.

25. Eliade, *Shamanism*, 190–200.

26. Eliade, *Myths, Dreams, and Mysteries*, 65.

27. Ibid., 64.

28. Morris Berman, *Coming to Our Senses* (New York: Bantam, 1989), 68.

29. Ibid., 69.

30. Gebser, *The Ever-Present Origin*, 49.

31. Ibid., 52.

32. Ibid., 51.

33. Ibid., 121.

34. Ibid., 55.

35. For references to impulse and instinct, see ibid., 46, 60, 67; for the relationship to sexuality, see ibid., 208, 230.

36. Ibid., 270.

37. Ibid., 145.

38. Ibid. (italics mine).

39. Ibid., 46.

40. Jules Lachelier, "Psychology and Metaphysics," in *The Search for Being*, ed. and trans. Jean T. Wilde and William Kimmel (New York: Noonday Press, 1962); Lachelier discusses the dimensional hierarchy on 166–70, and the hierarchy of nature on 154–55.

41. Ibid., 172.

42. Ibid., 168.

43. Ibid., 155.

44. P. D. Ouspensky, *Tertium Organum* (New York: Vintage, 1970), 74.

45. Ibid., 77–78.

46. Gebser, *The Ever-Present Origin*, 121.

47. Ibid., 49, 48.

48. The aurochs, for example, was a type of wild cattle (now extinct) that was worshipped as a divine figure during the Neolithic Age.

49. Selma Fraiberg, *The Magic Years* (New York: Charles Scribner's Sons, 1959), 16–23.

50. "Self-domestication," Wikipedia, accessed September 5, 2013, at http://en.wikipedia.org/w/index.php?title=Self-domestication&oldid=546618587.

51. Raymond Coppinger and Lorna Coppinger, *Dogs: A New Understanding of Canine Origin, Behavior and Evolution* (Chicago: University of Chicago Press, 2002), 61.

52. Eliade, *Myths, Dreams, and Mysteries*, 59.

53. Eliade, *Shamanism*, 476–77.

54. Ibid., 71–81.

55. Ibid., 72–73.

56. Ibid., 24ff.

57. Ibid., 476–77.

58. Joseph Shipley, *The Origins of English Words* (Baltimore: Johns Hopkins University Press, 1984), 70.

59. Ibid.

60. Eliade, *Shamanism*, 29.

61. Ibid.

62. Ibid., 30.

63. For readers not familiar with the psychological literature, the bracketed phrase is a variation on the term "regression in the service of the ego." This expression refers to the deliberate induction of an inchoate state of consciousness in the interest of the mature ego, as is done in the work of artists, creative writers, and the like.

64. Beginning with Proust's famous account of how a simple aroma unleashed a cascade of memories from early childhood, there has been a wealth of anecdotal data suggesting that olfactory cues can have such an effect. The experimental research of Rachel Herz seems to lend credence to these reports; see Rachel Sarah Herz, "Are Odors the Best Cues to Memory? A Cross-Modal Comparison of Associative Memory Stimuli," *Annals of the New York Academy of Sciences* 855 (1998): 670–74. See also Michael Stoddard, *The Scented Ape* (Cambridge, UK: Cambridge University Press, 1990).

65. Their synchronous spinning is in pairs, as specified at the end of the previous section.

Chapter 7

1. C. G. Jung, *Mysterium Coniunctionis*, vol. 14, *The Collected Works of C. G. Jung*, trans. R. F. C. Hull (Princeton, NJ: Princeton University Press, 1970), 534.

2. Ibid., 138.

3. Ibid.

4. Jung's reference to Zen appears in ibid., 537.

5. Hajime Tanabe, *Philosophy as Metanoetics*, trans. Takeuchi Yoshinori (Berkeley: University of California Press, 1986), 56.

6. Ibid., 7–12.

7. Ibid., 22.

8. Ibid., 18, 74.

9. Ibid., 85.

10. Ibid., 56.

11. C. G. Jung, *Alchemical Studies*, vol. 13, *The Collected Works of C. G. Jung*, trans. R. F. C. Hull (Princeton, NJ: Princeton University Press, 1967), 96.

12. Ibid., 238.

13. Ibid., 239.

14. C. G. Jung, *Psychology and Alchemy*, vol. 12, *The Collected Works of C. G. Jung*, trans. R. F. C. Hull (Princeton, NJ: Princeton University Press, 1968), 304.

15. Ibid.

16. C. G. Jung, *Psychological Types*, vol. 6, *The Collected Works of C. G. Jung*, trans. R. F. C. Hull (Princeton, NJ: Princeton University Press, 1971), 519.

17. Jean Gebser, *The Ever-Present Origin* (Athens: Ohio University Press, 1985); the reference to "deep sleep" appears on 121; the remainder of the citation is found on 44–45.

18. Eugene T. Gendlin, "Thinking Beyond Patterns: Body, Language, and Situations," in *The Presence of Feeling in Thought*, ed. Bernard den Ouden and Marcia Moen (New York: Peter Lang, 1991), 61.

19. Ibid., 63.

20. Ibid., 81.

21. Ibid., 75.

22. Jung, *Mysterium Coniunctionis*, 45–46.

23. Ibid., 48–49.

24. Ibid., 49.

25. Ibid., 51.

26. Ibid., 52.

27. Ibid., 53.

28. Ibid., 56.

29. Paul D. MacLean, *The Triune Brain in Evolution* (New York: Plenum, 1990), 23.

30. Erich Jantsch, *The Self-Organizing Universe* (New York: Pergamon, 1980), 165–66.

31. Ibid., 165.

32. Discussion of the "alimentary uroboros" appears in Erich Neumann, *The Origins and History of Consciousness* (Princeton, NJ: Princeton University Press, 1954), 27–34.

33. Jung, *Mysterium Coniunctionis*, 185.

34. Neumann, *The Origins and History of Consciousness*, 26–27.

35. Ibid., 28.

36. Ibid., 29.

37. Ibid., 29–30.

38. Ibid., 30.

39. Ibid., 290–91. As discussed in chapter 1, Neumann defined "centroversion" as the process through which consciousness achieves a well-formed and stable ego serving as the center of an individualized identity.

40. Seymour Boorstein, *Transpersonal Psychotherapy* (Albany: State University of New York Press, 1996), 365.

41. Diego L. Rapoport, "Surmounting the Cartesian Cut Further," in *Focus on Quantum Mechanics*, ed. David E. Hathaway and Elizabeth M. Randolf (Hauppauge, NY: Nova Science Publishers, 2012).

42. Mladen J. Kocica et al., "The Helical Ventricular Myocardial Band," *European Journal of Cardio-Thoracic Surgery* 29 (2006): S21–S40.

43. James Joyce, *Ulysses* (1922; repr., New York: Random House, 1986).

44. Christine van Boheemen-Saaf, "Shape and Satisfaction: The Figure of the Aged Penelope in Dickens and Joyce," *Papers on Joyce* 10/11 (2004–2005): 47.

45. Ibid., 52.

46. Jacqueline Kay Thomas, "Aphrodite Unshamed: James Joyce's Romantic Aesthetics of Feminine Flow" (PhD dissertation, University of Texas at Austin, 2007), 7.

47. Ibid., 11.

48. Ibid., 125.

49. Michael Stanier, "'The Void Awaits Surely All Them That Weave the Wind': 'Penelope' and 'Sirens' in *Ulysses*," *Twentieth Century Literature* 41, no. 3 (1995): 330.

50. Robert Boyle, "Penelope," in *James Joyce's Ulysses*, ed. Clive Hart and David Hayman (Berkeley: University of California Press, 1977), 412.

51. Rory Goff, website, accessed August 2, 2011, at http://www.artesmagicae.com/VesicaPiscis.htm.

52. Lawrence Blair, *Rhythms of Vision* (New York: Schocken, 1975), 75.

53. Buffie Johnson, *Lady of the Beasts: The Goddess and Her Sacred Animals* (Vermont: Inner Traditions, 1994), 224.

54. Olive Whicher, *Projective Geometry: Creative Polarities in Space and Time* (London: Rudolf Steiner, 1971), 270.

55. This 1876 illustration of Mercurius is from Ernst Wallis, *Illustrerad Verldhistoria* (Chicago: Svenska Amerikanaren, 1895).

56. This image appears on a twenty-first-century BCE commemorative vase, the "libation vase of Gudea." The deity is surrounded by griffins, powerful mythological creatures who serve as its guardians.

57. Walter J. Friedlander, *The Golden Wand of Medicine* (Westport, CT: Greenwood Press, 1992), 17.

58. Ibid., 21.

59. Demetrios Vakras, "Origins of Snake Worship," accessed August 14, 2011, at http://www.daimonas.com/pages/snake-worship.html; Van James, *Spirit and Art: Pictures of the Transformation of Consciousness* (Great Barrington, MA: Anthroposophic Press, 2001), 253.

60. Linda Foubister, *Goddess in the Grass: Serpentine Mythology and the Great Goddess* (Victoria, BC: Eccenova Editions, 2003), 7.

61. François P. Retief and Louise Cilliers, "Snake and Staff Symbolism and Healing," *Acta Theologica Supplementum* 7 (2005): 189.

62. William Irwin Thompson, *The Time Falling Bodies Take to Light* (New York: Palgrave Macmillan, 1981), 111–12.

63. Ibid., 112.

64. Ibid., 113–14.

65. Ibid., 111.

66. Ibid.

67. Monica Sjöö and Barbara Mor, *The Great Cosmic Mother* (New York: Harper & Row, 1987), 46.

68. Ibid., 48–49.

69. Ibid., 60.

70. György Doczi, *The Power of Limits* (Boulder, CO: Shambhala, 1981), 27.

71. Ibid., 27–28.

72. Ibid., 26.

73. Roger Lewin and Robert Foley, *Principles of Human Evolution* (Malden: MA: Blackwell, 2004), 485. There are indeed archaeologists who would question the idea that depictive symbolic representation is so scarce prior to the Upper Paleolithic. Given our purposes, however, it would take us too far afield to join a debate on the issue.

74. Ibid.

75. Paul G. Bahn and Jean Vertut, *Journey through the Ice Age* (Berkeley: University of California Press, 1997), 25.

76. Marija Gimbutas, *The Language of the Goddess* (New York: Harper & Row, 1989), 19.

77. Ibid., 19, 121.

78. See Balaji Mundkur, *The Cult of the Serpent* (Albany: State University of New York Press, 1983); Foubister, *Goddess in the Grass*; Bruno Barbatti, *Berber Carpets of Morocco* (Paris: ACR, 2008).

79. Thompson, *The Time Falling Bodies Take to Light*, 111.

80. Ibid., 112.

81. Joseph Campbell, *The Mythic Image* (Princeton, NJ: Princeton University Press, 1981), 283.

82. C. G. Jung, *Visions* (Princeton, NJ: Princeton University Press, 1997), 598–99.

83. Ibid., 601.

84. Herbert V. Guenther, foreword to Swami Sivananda Radha, *Kundalini, Yoga for the West* (Spokane, WA: Timeless Books, 1978), xviii.

85. John C. Huntington and Dina Bangdel, *The Circle of Bliss* (Chicago: Serindia Publications, 2003), 250.

86. Nahum Stiskin, *The Looking-Glass God* (New York: Autumn Press, 1972), 90.

87. Rob Preece, *The Psychology of Buddhist Tantra* (Ithaca, NY: Snow Lion Publications, 2006), 215.

88. Ibid., 218, 228.

89. Ibid., 224.

90. Ibid., 230.

91. Ibid., 216.

92. Ibid., 225–226.

93. Ibid., 227.

94. Ibid., 171.

95. Stanislav Grof, "Modern Consciousness Research and Human Survival," *Re-Vision* 8, no. 1 (1985): 27–39.

96. Ibid., 30.

97. Ibid., 30–31.

98. Ibid., 37.

99. Ibid.

100. Ibid.

101. Ibid., 28.

102. Ibid.

103. Ibid., 34.
104. Ibid.
105. Ibid., 27.
106. Ibid., 36.
107. Ibid., 31.
108. David R. Kinsley, *Hindu Goddesses* (Berkeley: University of California Press, 1988).

Abbott, Edwin. *Flatland: A Romance of Many Dimensions*. 1884. Reprint, New York: Barnes & Noble, 1983.

Abram, David. *The Spell of the Sensuous*. New York: Vintage, 1996.

Applebaum, David. "On Turning a Zen Ear." *Philosophy East and West* 33 (1983): 115–21.

Baars, Bernard J., and Nicole M. Gage. *Cognition, Brain, and Consciousness: Introduction to Cognitive Neuroscience*. Burlington, MA: Academic Press, 2010.

Bahn, Paul G., and Jean Vertut. *Journey through the Ice Age*. Berkeley: University of California Press, 1997.

Barbatti, Bruno. *Berber Carpets of Morocco*. Paris: ACR, 2008.

Barfield, Owen. *Saving of Appearances*. Middletown, CT: Wesleyan University Press, 1988.

Becker, Ernest. *The Denial of Death*. New York: Free Press, 1973.

Berger, John, and Jean Mohr. *Another Way of Telling*. New York: Pantheon Books, 1982.

Berman, Morris. *Coming to Our Senses*. New York: Bantam, 1989.

Blair, Lawrence. *Rhythms of Vision*. New York: Schocken, 1975.

Bohm, David. "The Bohm/Rosen Correspondence." In *Science, Paradox, and the Moebius Principle*, edited by Steven M. Rosen, 223–58. Albany: State University of New York Press, 1994.

Boorstein, Seymour. *Transpersonal Psychotherapy*. Albany: State University of New York Press, 1996.

Boyle, Robert. "Penelope." In *James Joyce's Ulysses*, edited by Clive Hart and David Hayman, 407–35. Berkeley: University of California Press, 1977.

Braine, Susan. *Drumbeat/Heartbeat: A Celebration of the Powwow*. Minneapolis: Lerner Publications, 1995.

Braun, Allen R., T. J. Balkin, N. J. Wesenten, R. E. Carson, M. Varga, P. Baldwin, S. Selbie, G. Belenky, and P. Herscovitch. "Regional Cerebral Blood Flow Throughout the Sleep-Wake Cycle." *Brain* 120 (1997): 1173–97.

Braun, Allen R., T. J. Balkin, N. J. Wesenten, F. Gawdry, R. E. Carson, M. Varga, P. Baldwin, G. Belenky, and P. Herscovitch. "Dissociated Pattern of Activity in

Visual Cortices and Their Projections During Human Rapid Eye Movement Sleep." *Science* 279 (1998): 91–95.

Burrow, Trigant. *Science and Man's Behavior.* New York: Philosophical Library, 1953.

Campbell, Joseph. *The Mythic Image.* Princeton, NJ: Princeton University Press, 1981.

Conger, John P. *Jung and Reich: The Body as Shadow.* Berkeley, CA: North Atlantic Books, 1988.

Coppinger, Raymond, and Lorna Coppinger. *Dogs: A New Understanding of Canine Origin, Behavior and Evolution.* Chicago: University of Chicago Press, 2002.

Couliano, I. *Eros and Magic in the Renaissance.* Chicago: University of Chicago Press, 1987.

Deleuze, Gilles, and Felix Guattari. *A Thousand Plateaus: Capitalism and Schizophrenia.* Minneapolis: University of Minnesota Press, 1987.

Doczi, György. *The Power of Limits.* Boulder, CO: Shambhala, 1981.

Eliade, Mircea. *Myths, Dreams, and Mysteries.* New York: Harper & Row, 1960.

———. *The Forge and the Crucible.* New York: Harper & Row, 1962.

———. *Shamanism.* Princeton, NJ: Princeton University Press, 1964.

Escher, M. C. *The Graphic Work of M. C. Escher.* New York: Ballantine, 1971.

Flew, Antony. *A Dictionary of Philosophy.* New York: St. Martin's Press, 1979.

Foubister, Linda. *Goddess in the Grass: Serpentine Mythology and the Great Goddess.* Victoria, BC: Eccenova Editions, 2003.

Fraiberg, Selma. *The Magic Years.* New York: Charles Scribner's Sons, 1959.

Friedlander, Walter J. *The Golden Wand of Medicine.* Westport, CT: Greenwood Press, 1992.

Galt, Alfreda. "Trigant Burrow and the Laboratory of the 'I.'" *The Humanistic Psychologist* 23 (1995): 19–39.

Gardner, Martin. *The Ambidextrous Universe.* New York: Charles Scribner's Sons, 1979.

Gebser, Jean. *The Ever-Present Origin.* Athens: Ohio University Press, 1985.

Gendlin, Eugene T. *Focusing.* New York: Bantam, 1978.

———. "Thinking Beyond Patterns: Body, Language, and Situations." In *The Presence of Feeling in Thought,* edited by Bernard den Ouden and Marcia Moen, 27–189. New York: Peter Lang, 1991.

———. "Words Can Say How They Work." In *Proceedings, Heidegger Conference,* edited by Robert P. Crease, 29–35. Stony Brook: Department of Philosophy, State University of New York at Stony Brook, 1993.

Gimbutas, Marija. *The Language of the Goddess.* New York: Harper & Row, 1989.

Grof, Stanislav. "Modern Consciousness Research and Human Survival." *Re-Vision* 8, no. 1 (1985): 27–39.

Guenther, Herbert V. Foreword to Swami Sivananda Radha, *Kundalini, Yoga for the West.* Spokane, WA: Timeless Books, 1978.

Hammer, Anita. "Mirroring and the Topology of Theatre." Presented in Cross-Disciplinary Seminar, Department of Social Anthropology, Norwegian University of Science and Technology, Trondheim, Norway, February 15, 1999.

Heidegger, Martin. "Modern Science, Metaphysics, and Mathematics" (1962). In *Martin Heidegger: Basic Writings,* edited by David F. Krell, 247–82. New York: Harper & Row, 1977.

———. "The End of Philosophy and the Task of Thinking" (1964). In *Martin Heidegger: Basic Writings,* edited by David F. Krell, 373–92. New York: Harper & Row, 1977.

Herz, Rachel Sarah. "Are Odors the Best Cues to Memory? A Cross-Modal Comparison of Associative Memory Stimuli." *Annals of the New York Academy of Sciences* 855 (1998): 670–74.

Hillman, James. *The Dream and the Underworld.* New York: Harper & Row, 1979.

Holton, Gerald. *Thematic Origins of Scientific Thought.* Cambridge, MA: Harvard University Press, 1988.

Huntington, John C., and Dina Bangdel. *The Circle of Bliss.* Chicago: Serindia Publications, 2003.

Jaffé, Aniela. *From the Life and Work of C. G. Jung.* Einsiedeln, Switzerland: Daimon, 1989.

James, Van. *Spirit and Art: Pictures of the Transformation of Consciousness.* Great Barrington, MA: Anthroposophic Press, 2001.

Jantsch, Erich. *The Self-Organizing Universe.* New York: Pergamon, 1980.

Jaynes, Julian. *The Origin of Consciousness in the Breakdown of the Bicameral Mind.* Boston: Houghton Mifflin, 1976.

Johnson, Buffie. *Lady of the Beasts: The Goddess and Her Sacred Animals.* Vermont: Inner Traditions, 1994.

Jones, W. T. *A History of Western Philosophy.* New York: Harcourt, Brace & World, 1952.

Joyce, James. *Ulysses.* 1922. Reprint, New York: Random House, 1986.

Jung, C. G. *Aion,* vol. 9ii, *The Collected Works of C. G. Jung.* Translated by R. F. C. Hull. Princeton, NJ: Princeton University Press, 1959.

———. *Alchemical Studies,* vol. 13, *The Collected Works of C. G. Jung.* Translated by R. F. C. Hull. Princeton, NJ: Princeton University Press, 1967.

———. *Psychology and Alchemy,* vol. 12, *The Collected Works of C. G. Jung.* Translated by R. F. C. Hull. Princeton, NJ: Princeton University Press, 1968.

———. *Mysterium Coniunctionis,* vol. 14, *The Collected Works of C. G. Jung.* Translated by R. F. C. Hull. Princeton, NJ: Princeton University Press, 1970.

———. *Psychological Types,* vol. 6, *The Collected Works of C. G. Jung.* Translated by R. F. C. Hull. Princeton, NJ: Princeton University Press, 1971.

———. *Memories, Dreams, Reflections.* New York: Vintage, 1989.

———. *Visions.* Princeton, NJ: Princeton University Press, 1997.

Kierkegaard, Søren. "Truth Is Subjectivity" (1846). In *Existentialism from Dostoevsky to Sartre,* edited by Walter Kaufmann, 110–20. New York: New American Library, 1975.

Kinsley, David R. *Hindu Goddesses.* Berkeley: University of California Press, 1988.

Kittelson, Mary Lynn. "The Acoustic Vessel." In *The Interactive Field in Analysis,* edited by Murray Stein, 89–105. Wilmette, IL: Chiron, 1995.

Kocica, Mladen J., A. F. Corno, V. Lackovic, and V. I. Kanjuh. "The Helical Ventricular Myocardial Band." *European Journal of Cardio-Thoracic Surgery* 29 (2006): S21–S40.

Lacan, Jacques. "Of Structure as an Inmixing of an Otherness Prerequisite to Any Subject Whatever" (1966). In *The Languages of Criticism and the Sciences of Man: The Structuralist Controversy,* edited by Richard Macksey and Eugenio Donato, 186–200. Baltimore: Johns Hopkins University Press, 1970.

Lachelier, Jules. "Psychology and Metaphysics." In *The Search for Being,* edited and translated by Jean T. Wilde and William Kimmel, 153–69. New York: Noonday Press, 1962.

Lavery, David. "The Eye of Longing." *Re-Vision* 6, no. 1 (1983): 22–33.
Leder, Drew. *The Absent Body*. Chicago: University of Chicago Press, 1990.
Lewin, Roger, and Robert Foley. *Principles of Human Evolution*. Malden, MA: Blackwell, 2004.
MacLean, Paul D. "Alternative Neural Pathways to Violence." In *Alternatives to Violence*, edited by L. Ng, 22–34. Alexandria, VA: Time-Life, 1968.
———. *The Triune Brain in Evolution*. New York: Plenum, 1990.
Mandelbrot, Benoit. *Fractals*. San Francisco: Freeman, 1977.
Maquet, Pierre, J. Péters, J. Aerts, G. Delfiore, C. Dequeldre, A. Luxen, and G. Franck. "Functional Neuroanatomy of Human Rapid-Eye-Movement Sleep and Dreaming." *Nature* 383 (1996): 163–66.
Merleau-Ponty, Maurice. *Phenomenology of Perception*. Translated by Colin Smith. London: Routledge and Kegan Paul, 1962.
———. *The Visible and the Invisible*. Translated by Alphonso Lingis. Evanston, IL: Northwestern University Press, 1968.
Metzner, Ralph. *Maps of Consciousness*. New York: Macmillan, 1971.
Mundkur, Balaji. *The Cult of the Serpent*. Albany: State University of New York Press, 1983.
Neumann, Erich. *The Origins and History of Consciousness*. Princeton, NJ: Princeton University Press, 1954.
Nietzsche, Friedrich. "On the Vision and the Riddle" (1878). In *The Portable Nietzsche*, edited and translated by Walter Kaufmann, 267–72. New York: Viking, 1968.
Obrist, Barbara. "Visualization in Medieval Alchemy." *HYLE—International Journal for Philosophy of Chemistry* 9, no. 2 (2003): 131–70.
Ong, Walter. *Interfaces of the Word*. Ithaca, NY: Cornell University Press, 1977.
Ouspensky, P. D. *Tertium Organum*. New York: Vintage, 1970.
Peirce, Charles S. *Collected Papers–II*. Edited by Charles Hartshorne and Paul Weiss. Cambridge, MA: Harvard University Press, 1933.
Preece, Rob. *The Psychology of Buddhist Tantra*. Ithaca, NY: Snow Lion Publications, 2006.
Rapoport, Diego L. "Surmounting the Cartesian Cut through Philosophy, Physics, Logic, Cybernetics, and Geometry." *Foundations of Physics* 41 (2011): 33–76.
———. "Surmounting the Cartesian Cut Further." In *Focus on Quantum Mechanics*, edited by David E. Hathaway and Elizabeth M. Randolf. Hauppauge, NY: Nova Science Publishers, 2012.
Read, John. *Prelude to Chemistry*. Cambridge, MA: MIT Press, 1966.
Retief, François P., and Louise Cilliers. "Snake and Staff Symbolism and Healing." *Acta Theologica Supplementum* 7 (2005): 189–99.
Rosen, Steven M. "A Plea for the Possibility of Visualizing Existence." *Scientia* 108, nos. 9–12 (1973): 789–802.
———. "A Neo-Intuitive Proposal for Kaluza-Klein Unification." *Foundations of Physics* 18, no. 11 (1988): 1093–1139.
———. *Science, Paradox, and the Moebius Principle*. Albany: State University of New York Press, 1994.
———. "Pouring Old Wine into a New Bottle." In *The Interactive Field in Analysis*, edited by Murray Stein, 121–41. Wilmette, IL: Chiron, 1995.
———. *Dimensions of Apeiron*. Amsterdam: Editions Rodopi, 2004.
———. *Topologies of the Flesh*. Athens: Ohio University Press, 2006.

———. *The Self-Evolving Cosmos*. London: World Scientific Publishing, 2008.

Rucker, Rudolph. *Geometry, Relativity, and the Fourth Dimension*. New York: Dover, 1977.

Ryan, Paul. *Video Mind/Earth Mind*. New York: Peter Lang, 1993.

Schwartz-Salant, Nathan. *The Mystery of Human Relationship*. London: Routledge, 1998.

———. *The Black Nightgown*. Wilmette, IL: Chiron, 2007.

Sheets-Johnstone, Maxine. *The Roots of Thinking*. Philadelphia: Temple University Press, 1990.

Shipley, Joseph. *The Origins of English Words*. Baltimore: Johns Hopkins University Press, 1984.

Sjöö, Monica, and Barbara Mor. *The Great Cosmic Mother*. New York: Harper & Row, 1987.

Sloan, Lisa. "Shamanic Initiation: Map of the Soul." Doctoral dissertation, Pacifica Graduate Institute, 1999.

Spiegelberg, Herbert. *The Phenomenological Movement*. The Hague: Martinus Nijhoff, 1982.

Spivak, Gayatri C. "Translator's Preface." In Jacques Derrida, *Of Grammatology*, ix–lxxxvii. Baltimore: Johns Hopkins University Press, 1976.

Stanier, Michael. "'The Void Awaits Surely All Them That Weave the Wind': 'Penelope' and 'Sirens' in *Ulysses*." *Twentieth Century Literature* 41, no. 3 (1995): 319–31.

Steiner, Rudolf. *An Outline of Occult Science*. Spring Valley, NY: Anthroposophic Press, 1972.

Stiskin, Nahum. *The Looking-Glass God*. New York: Autumn Press, 1972.

Stoddard, Michael. *The Scented Ape*. Cambridge, UK: Cambridge University Press, 1990.

Tanabe, Hajime. *Philosophy as Metanoetics*. Translated by Takeuchi Yoshinori. Berkeley: University of California Press, 1986.

Thomas, Jacqueline Kay. "Aphrodite Unshamed: James Joyce's Romantic Aesthetics of Feminine Flow." PhD dissertation, University of Texas at Austin, 2007.

Thompson, William Irwin. *The Time Falling Bodies Take to Light*. New York: Palgrave Macmillan, 1981.

van Boheemen-Saaf, Christine. "Shape and Satisfaction: The Figure of the Aged Penelope in Dickens and Joyce." *Papers on Joyce* 10/11 (2004–2005): 45–56.

Vitz, Paul C., and Arnold B. Glimcher. *Modern Art and Modern Science*. New York: Praeger Books, 1984.

von Franz, Marie-Louise. *Number and Time*. Evanston, IL: Northwestern University Press, 1974.

———. "Psyche and Matter in Alchemy and Modern Science." *Quadrant* 8 (1975): 33–49.

———. *On Dreams and Death*. Boston: Shambhala, 1987.

Wallis, Ernst. *Illustrerad Verldhistoria*. Chicago: Svenska Amerikanaren, 1895.

Washburn, Michael. *The Ego and the Dynamic Ground*. Albany: State University of New York Press, 1988.

Whicher, Olive. *Projective Geometry: Creative Polarities in Space and Time*. London: Rudolf Steiner, 1971.

INDEX

Page numbers in italics indicate illustrations.